Emerging Trends in Beverage Processing

Emerging Trends in Beverage Processing

Editor

Antonio Morata

MDPI • Basel • Beijing • Wuhan • Barcelona • Belgrade • Manchester • Tokyo • Cluj • Tianjin

Editor
Antonio Morata
Universidad Politécnica de Madrid (UPM)
Spain

Editorial Office
MDPI
St. Alban-Anlage 66
4052 Basel, Switzerland

This is a reprint of articles from the Special Issue published online in the open access journal *Beverages* (ISSN 2306-5710) (available at: https://www.mdpi.com/journal/beverages/special_issues/processing_beverages).

For citation purposes, cite each article independently as indicated on the article page online and as indicated below:

LastName, A.A.; LastName, B.B.; LastName, C.C. Article Title. *Journal Name* **Year**, *Volume Number*, Page Range.

ISBN 978-3-0365-1090-3 (Hbk)
ISBN 978-3-0365-1091-0 (PDF)

Cover image courtesy of Dr. Carlos Escott.

Contents

About the Editor

Antonio Morata is a professor of Food Science and Technology at the Universidad Politécnica de Madrid (UPM), Spain, specializing in wine technology. He is the coordinator of the Master in Food Engineering Program at UPM, and a professor of enology and wine technology at the European Master of Viticulture and Enology, Euromaster Vinifera-Erasmus+. He is the Spanish delegate at the group of experts in wine microbiology and wine technology of the International Organisation of Vine and Wine (OIV). He is the author of more than 70 research articles, 3 books, 4 edited books, 6 special issues and 16 book chapters. He is also an Editorial Board Member of Fermentation MDPI and Associated Editor of Beverages MDPI.

 *beverages*

Editorial

Emerging Trends in Beverage Processing

Antonio Morata

enotecUPM, Chemistry and Food Technology Department, Escuela Técnica Superior de Ingeniería Agronómica, Alimentaria y de Biosistemas, Universidad Politécnica de Madrid, Avenida Puerta de Hierro 2, 28040 Madrid, Spain; antonio.morata@upm.es

Citation: Morata, A. Emerging Trends in Beverage Processing. *Beverages* **2021**, *7*, 8. https://doi.org/10.3390/beverages7010008

Received: 19 January 2021
Accepted: 25 January 2021
Published: 28 January 2021

Beverage processing is open to new technologies; among them, nonthermal physical technologies such as discontinuous hydrostatic pressure (HHP), ultrahigh-pressure homogenization (UHPH), pulsed electric field (PEF), ultrasound (US), atmospheric pressure cold plasma (APCP), or pulsed light (PL) are growing increasingly in the food industry. The potentiality to speed up the production process, to improve the quality, to develop new beverages or new features in conventional beverages, to reach more stable beverages with better safety, and to protect sensory and nutritional quality are key parameters allowed by these technologies. Additionally, emerging fermentation biotechnologies or new perspectives on sensory attributes also contribute to the development of new beverages with increased acceptance by consumers.

The use of emerging technologies in the processing of fruits and juices for beverage production was strongly developed in the last 20 years. Technologies such as HHP, UHPH, PEF, US, APCP, or PL are used to process fruits or juices, increasing extraction yields, inactivating microorganism, improving colloidal stability, enhancing long term microbial and physicochemical stability [1], and additionally allowing the use of new fermentation biotechnologies [2] with positive sensory impacts [2,3]. Discontinuous hydrostatic pressure (HHP) facilitates better extraction and microbial control, preserving juice quality [4], and can be used as a storage technology to keep stable beverage properties at room temperature [5]. New materials will allow for the optimization of the problems concerning vessels for high-pressure storage. Continuous (ultra)high-pressure homogenization ((U)HPH) is an efficient and highly reliable technology for microbial control and even sterilization depending on in-valve temperatures with gentle management of the nutritional and sensory features [6,7], allowing for even effective destruction of oxidative enzymes (PPO) and therefore minimizing the use of chemical additives like sulfites [8]. The use of pulsed electric fields (PEF) and ultrasounds (US) facilitates the extraction of pigments and tannins, facilitating winemaking technologies [9], controlling microorganisms, and improving fermentation biotechnologies such as the use of non-*Saccharomyces* yeasts [10]. PEF extraction and microbial control are due to cell electroporation, and ultrasound extraction is enabled by cavitation phenomena [9]. Atmospheric pressure cold plasma (APCP) is a nonthermal technology that promotes higher color intensity and more tannins, improving wine quality [11]. Plasma is a gas state that contains ionized particles that can be applied in beverages directly or indirectly to process liquids and to control microorganisms. Pulsed light (PL) is the use of high-intensity UV-visible-IR radiation during an ultrashort flash to eliminate microorganisms in fruits and in juices or beverages [12,13]. PL can be considered a nonthermal technology producing temperature increments no higher than 3 °C [12]. Some effects on extraction and enzyme control have been described [12]. The effectivity is similar to or higher than UV continuous radiation but in a faster and more effective process. Microbial control by emerging nonthermal technologies facilitate the implantation of new fermentation biotechnologies as the use of non-*Saccharomyces* yeasts [2] or yeast–bacteria co-inoculations with *Lactobacillus plantarum* [14]. *L. plantarum* opens new possibilities in the control of malolactic fermentation in wines. Finally, sensory assessment is a hot topic in the evaluation of quality in beverages, though some confusing attributes are frequently used to describe high-quality beverages as wines. Minerality is a complex concept that is used to describe premium wines that are connected to production areas and reflect stony and mineral perceptions. It has been observed that, in fact, such sensory perceptions may not be correlated with mineral compounds and must be understood as sensory impacts produced by organic volatile compounds [15].

Author Contributions: Conceptualization and writing—review and editing, A.M. The author have read and agreed to the published version of the manuscript.

Funding: This research received no external funding.

Institutional Review Board Statement: Not applicable.

Informed Consent Statement: Not applicable.

Conflicts of Interest: The authors declare no conflict of interest.

References

1. Morata, A.; Loira, I.; Vejarano, R.; González, C.; Callejo, M.J.; Suárez-Lepe, J.A. Emerging preservation technologies in grapes for winemaking. *Trends Food Sci. Technol.* **2017**, *67*, 36–43. [CrossRef]
2. Morata, A.; Escott, C.; Bañuelos, M.A.; Loira, I.; del Fresno, J.M.; González, C.; Suárez-Lepe, J.A. Contribution of Non-*Saccharomyces* Yeasts to Wine Freshness. A Review. *Biomolecules* **2020**, *10*, 34. [CrossRef] [PubMed]
3. Morata, A.; Escott, C.; Loira, I.; Del Fresno, J.M.; González, C.; Suárez-Lepe, J.A. Influence of *Saccharomyces* and non-*Saccharomyces* Yeasts in the Formation of Pyranoanthocyanins and Polymeric Pigments during Red Wine Making. *Molecules* **2019**, *24*, 4490. [CrossRef] [PubMed]
4. Morata, A.; Loira, I.; Vejarano, R.; Bañuelos, M.A.; Sanz, P.D.; Otero, L.; Suárez-Lepe, J.A. Grape Processing by High Hydrostatic Pressure: Effect on Microbial Populations, Phenol Extraction and Wine Quality. *Food Bioprocess. Technol.* **2015**, *8*, 277–286. [CrossRef]
5. Otero, L. Hyperbaric Storage at Room Temperature for Fruit Juice Preservation. *Beverages* **2019**, *5*, 49. [CrossRef]
6. Comuzzo, P.; Calligaris, S. Potential Applications of High Pressure Homogenization in Winemaking: A Review. *Beverages* **2019**, *5*, 56. [CrossRef]
7. Morata, A.; Guamis, B. Use of UHPH to obtain juices with better nutritional quality and healthier wines with low levels of SO_2 MINIREVIEW. *Front. Nutr.* **2020**, *7*, 598286. [CrossRef] [PubMed]
8. Bañuelos, M.A.; Loira, I.; Guamis, B.; Escott, C.; Del Fresno, J.M.; Codina-Torrella, I.; Quevedo, J.M.; Gervilla, R.; Rodríguez Chavarría, J.M.; de Lamo, S.; et al. White wine processing by UHPH without SO_2. Elimination of microbial populations and effect in oxidative enzymes, colloidal stability and sensory quality. *Food Chem.* **2020**, *332*, 127417. [CrossRef]
9. Maza, M.; Álvarez, I.; Raso, J. Thermal and Non-Thermal Physical Methods for Improving Polyphenol Extraction in Red Winemaking. *Beverages* **2019**, *5*, 47. [CrossRef]
10. Morata, A.; Bañuelos, M.A.; Loira, I.; Raso, J.; Álvarez, I.; Garcíadeblas, B.; González, C.; Suárez Lepe, J.A. Grape Must Processed by Pulsed Electric Fields: Effect on the Inoculation and Development of Non-*Saccharomyces* Yeasts. *Food Bioprocess. Technol.* **2020**, *13*, 1087–1094. [CrossRef]
11. Sainz-García, E.; López-Alfaro, I.; Múgica-Vidal, R.; López, R.; Escribano-Viana, R.; Portu, J.; Alba-Elías, F.; González-Arenzana, L. Effect of the Atmospheric Pressure Cold Plasma Treatment on Tempranillo Red Wine Quality in Batch and Flow Systems. *Beverages* **2019**, *5*, 50. [CrossRef]
12. Santamera, A.; Escott, C.; Loira, I.; del Fresno, J.M.; González, C.; Morata, A. Pulsed Light: Challenges of a Non-Thermal Sanitation Technology in the Winemaking Industry. *Beverages* **2020**, *6*, 45. [CrossRef]
13. Escott, C.; Vaquero, C.; del Fresno, J.M.; Bañuelos, M.A.; Loira, I.A.; Han, S.; Bi, Y.; Morata, A.; Suárez-Lepe, J. Pulsed Light Effect in Red Grape Quality and Fermentation. *Food Bioprocess. Technol.* **2017**, *10*, 1540–1547. [CrossRef]
14. Krieger-Weber, S.; Heras, J.M.; Suarez, C. *Lactobacillus plantarum*, a New Biological Tool to Control Malolactic Fermentation: A Review and an Outlook. *Beverages* **2020**, *6*, 23. [CrossRef]
15. Zaldívar Santamaría, E.; Molina Dagá, D.; Palacios García, A.T. Statistical Modelization of the Descriptor "Minerality" Based on the Sensory Properties and Chemical Composition of Wine. *Beverages* **2019**, *5*, 66. [CrossRef]

Review

Hyperbaric Storage at Room Temperature for Fruit Juice Preservation

Laura Otero

MALTA Consolider Team, Institute of Food Science, Technology and Nutrition (ICTAN-CSIC) c/José Antonio Novais, 10, 28040 Madrid, Spain; l.otero@ictan.csic.es; Tel.: +34-91-544-5607

Received: 20 May 2019; Accepted: 4 July 2019; Published: 2 August 2019

Abstract: Hyperbaric storage is an innovative preservation method that consists of storing food under pressure, either at room or at low temperature, for time periods of days, weeks, or months. Recent scientific literature shows that hyperbaric storage at room temperature (HS-RT) could be an efficient method for fruit juice preservation. Depending on the level applied, pressure can inhibit and even inactivate the endogenous microflora of the fresh juice, while properly preserving other organoleptic and quality indicators. Even though the method has not yet been implemented in the food industry, its industrial viability has been evaluated from different points of view (product quality, consumer acceptation, vessel design, economic, or environmental, among others). The results reveal that HS-RT is effective in extending the shelf-life of both acidic and low-acidic fruit juices. Moreover, the energetic costs and the carbon footprint of HS-RT are considerably lower than those of refrigeration, therefore, HS-RT could be a reliable and environmentally friendly alternative to conventional cold storage. However, before industrial implementation, much more research is needed to clarify the effects of the storage conditions on the agents that cause fruit juice deterioration.

Keywords: hyperbaric storage; high pressure; food preservation; fruit juice

---

1. Introduction

The global fruit juice market reached a volume of 45.4 billion liters in 2018 and almost a quarter of this total volume was consumed in the EU [1]. Here, the per capita consumption of chilled juices has been continuously growing over the last five years, probably because many consumers perceive chilled products as healthier and more natural than ambient offerings [2]. However, the increasing demand of chilled products can have negative environmental consequences as food refrigeration is directly implicated in global warming and climate change [3,4]. It causes both direct greenhouse gas (GHG) emissions from the manufacture and direct loss of refrigerants as well as indirect GHG emissions from the energy use required to maintain the cold chain. Thus, food refrigeration is considered to be responsible for approximately 1% of the total global GHG emissions [5]. In the last decades, many strategies to reduce these emissions have been tested such as using refrigerants of low global warming potential, improving the insulation, design, and location of chilling and freezing facilities, or applying clean energy technologies, among others [5], but currently, food refrigeration still poses serious problems of sustainability. Therefore, to fight against climate change, novel, environmentally friendly storage methods with reduced energy requirements that are able to guarantee food safety and quality are urgently needed to replace refrigeration.

Recently, hyperbaric storage at room temperature (HS-RT) has been proposed as a potential alternative to refrigeration for food preservation. This novel storage method must not be confused with high-pressure processing that, nowadays, is a well-established technology in the food industry. In high-pressure processing, foods are treated at high pressure (about 300–700 MPa) for brief time periods, generally shorter than 10 min. The usual objective is to extend their shelf-life during the subsequent cold storage at atmospheric pressure. In contrast, in hyperbaric storage, the product is not treated at high pressure for some minutes, but stored under moderate pressure, usually not higher than 100 MPa, during the whole storage period, that is, days, weeks, or even months. The pursued aim is to use pressure as a limiting factor for food deterioration, just like low temperature in refrigeration. As food is preserved at room temperature, the only energy consumption is produced at the beginning of storage, during compression, because no additional energy is required to maintain the product under pressure for long times. Therefore, this system can involve considerable energy savings when compared with other food storage methods such as freezing or refrigeration. Obviously, hyperbaric storage can also be performed at low temperature, but in this case, energy costs would be increased.

Even though hyperbaric storage was first proposed as a food preservation method in 1977 [6], this technique has not really been explored until this century most likely because many doubts existed about its potential viability at an industrial scale. However, after the successful implementation of high-pressure processing in the food industry, the interest in hyperbaric storage has been revived. In 1997 and 2000, two patents were published [7,8] that described how a huge variety of raw and cooked foods and food ingredients could be preserved under pressure at 18–23 °C. Next, papers about hyperbaric storage at room temperature started to appear in the scientific literature, especially in this last decade. They proved that HS-RT was more efficient than conventional refrigeration for the preservation of not only fruit juices [9], but also meat [10–12] and milk products [13,14] and ready-to eat pre-cooked foods [15,16]. Despite this, HS-RT can be considered to be in its initial phase and, in fact, industrial equipment specifically designed for hyperbaric storage still has not been commercialized.

In this review, hyperbaric storage at room temperature is presented as a novel and environmentally friendly method for fruit juice preservation. Data from the existing literature have been gathered to critically evaluate the viability of the method from a multiple points of view, covering as many aspects as possible from juice safety and quality to consumer acceptability, equipment design, economic, and environmental issues. In this way, this review contributes to offering sustainable solutions to the food industry for fruit juice preservation.

2. Effect of Hyperbaric Storage on Agents Responsible for Juice Deterioration

It is obvious that the effectivity of HS-RT for fruit juice preservation depends on the effects that pressure produces on all of the agents responsible for juice deterioration. Among them, microbial agents are frequently considered the major concern, but enzymatic and chemical deterioration can also play a significant role in juice spoilage [17]. In the next subsections, the available information about the activity of all of these agents under pressure is presented. Microbial activity under pressure is assessed by comparing the microbial loads before and immediately after HS-RT, while enzymatic and chemical activities are evaluated by comparing the amount of either the reagents or the reaction products before and after HS-RT.

2.1. Microbial Load

Many papers in the literature have revealed that HS-RT is an effective method to extend the microbiological shelf-life of fruit juices over that achieved by conventional refrigeration [9,18–20]. For example, Lemos, Ribeiro, Fidalgo, Delgadillo, and Saraiva [18] reported that HS-RT at 75 MPa allowed for the shelf-life of watermelon juices to be extended for at least 21 days, while in conventional refrigeration, the product was completely spoilt after seven days. The effectiveness of HS-RT in hampering the cell-proliferating ability of microorganisms depends on several factors such as the storage pressure and time, the product characteristics, or the specific microorganisms present in the juice microbiota, among others.

The pressure level applied during hyperbaric storage plays a decisive role in the effects observed in the microflora of fruit juices. It is well-known that relatively low pressures (<50 MPa) are nonlethal for most mesophilic microorganisms, but affect some cellular processes such as motility, substrate transport, nutrient uptake, or cell division and growth [21–23]. However, most microorganisms can adapt themselves to these pressures and, although the lag phase frequently becomes longer as pressure increases, they can still proliferate [24]. When pressure is increased, DNA replication, translation, and transcription can be affected, and at 100–200 MPa and higher, microbial viability can be significantly reduced through multi-target inactivation mechanisms [21–23]. Therefore, depending on the pressure employed, HS-RT can not only slow down the microbial growth as refrigeration does, but also produces some damage in the microorganisms, resulting in microbial inactivation. For example, Bermejo-Prada et al. [25] observed that after one day of storage at 25 MPa, the initial counts of total aerobic mesophiles (TAM) and lactic acid bacteria (LAB) in strawberry juice remained almost invariable, while yeasts and molds (YM) decreased slightly. In contrast, storage at 50 MPa produced a slight decrease not only in the YM counts, but also in the TAM and LAB counts, while at 100 and 200 MPa, significant TAM, LAB, and YM reductions of 1.4, 1.6, and 1.0 $\log_{10}$ units and of 3.6, 3.6, and 3.1 $\log_{10}$ units were detected, respectively.

Moreover, microbial growth during hyperbaric storage is also affected by the storage time. In general, the longer the storage time, the greater the microbial damage produced [18,19,25,26]. Thus, Lemos, Ribeiro, Fidalgo, Delgadillo, and Saraiva [18] stored watermelon juices at 75 MPa and room temperature and, after three and 21 days of storage, they observed reductions close to 1 and 2 $\log_{10}$ units, respectively, in both the total aerobic mesophiles and total aerobic psychrophiles. However, it is important to note that at low storage pressures, microbial growth can occur after certain lag phase under pressure as previously mentioned. For example, Bermejo-Prada, López-Caballero, and Otero [25] observed that storage at 25 MPa for 1–10 days completely inhibited TAM, LAB, and YM growth in strawberry juice, but after 15 days of storage, the microbial load can be increased. Similar findings were described by Pinto, Moreira, Fidalgo, Santos, Vidal, Delgadillo, and Saraiva [19]. Thus, they stored watermelon juice at 50 MPa and room temperature and reported no significant increase in the total aerobic mesophiles and psychrophiles after four days of storage. However, after seven days, both counts increased by about 2 $\log_{10}$ units and reached values over the acceptability limit (>6 $\log_{10}$ CFU/mL). The elimination of the pressure labile microbiota at the beginning of the storage could favor the growth of pressure-resistant populations and contribute to the microbial growth observed at the end of storage in the above examples. Moreover, sublethal stresses could induce the expression of cell repair systems [27] and, therefore, an adaptation of some strains to stress could take place during long storage at relatively low pressures.

The intrinsic product characteristics are also a main factor that affects the effectiveness of HS-RT in preventing microbial spoilage. Some attributes such as the product composition, water activity, or pH can strongly impact on the effects observed after hyperbaric storage. For example, unlike in strawberry juice (pH = 3.3), HS-RT at 25 MPa did not slow down the growth of the natural microbiota in melon (pH = 5.7) and watermelon (pH = 5.8) juices [28,29], and pressures of 50 MPa and 75 MPa were needed to either reduce or completely stop microbial growth in these products [18,19,28,29]. The greater sensitivity to pressure observed in strawberry juice compared with melon and watermelon juices could be related to the low pH of this product. In this sense, Matsumura et al. [30] showed that pressure markedly narrowed the pH ranges for the growth of a variety of bacteria. Moreover, different authors in the literature have shown that as pH is lowered, most microbes become more susceptible to high-pressure inactivation [31–33].

Treatments applied to juices prior to HS-RT can also have an impact on microbial stability during hyperbaric storage. Thus, Segovia-Bravo, Guignon, Bermejo-Prada, Sanz and Otero [9] observed that the initial microbial load of frozen-thawed strawberry juice was reduced by more than 2 $\log_{10}$ units after 15 days of HS-RT at 25 MPa. At the end of storage, microbial levels were below the detection limit (10 CFU/mL for TAM and 100 CFU/mL for YM) and remained stable for at least 15 additional days at atmospheric pressure and 5 °C. In contrast, Bermejo-Prada, López-Caballero, and Otero [25] reported that in freshly squeezed strawberry juice, HS-RT at 25 MPa retarded microbial growth when compared to conventional cold storage, but could not completely avoid microbial proliferation. These results clearly indicate that microbial stability during HS-RT could be enhanced by the stress of the previous freeze–thaw treatment.

All the above results show that hyperbaric storage at room temperature can be an effective method to inhibit the growth of the endogenous microflora in fruit juices. To gain the first insight into the effects of HS-RT on pathogenic microorganisms, Pinto, Moreira, Fidalgo, Santos, Vidal, Delgadillo, and Saraiva [19] inoculated two specific microorganisms, *Listeria innocua* (ATCC 33090) and *Escherichia coli* (ATCC 25992), as surrogates for pathogen *L. monocytogenes* and pathogenic *E. coli*, in watermelon juice (3–4 $\log_{10}$ CFU/mL) and stored it for 10 days at 50, 75, or 100 MPa and room temperature. They found that after six days of storage at 50 MPa, *E. coli* counts were reduced below the detection limit (10 CFU/mL). However, this pressure was not enough to avoid the growth of *L. innocua* and storage at 75 MPa was needed to achieve the same results as in *E. coli*. These results confirm the greater sensitivity of Gram-negative bacteria (*E. coli*) to pressure when compared with Gram-positive bacteria (*L. innocua*), a well-known fact widely proven throughout the literature [32,34].

Furthermore, it is important to note that HS-RT can reduce not only the load of vegetative cells as conventional pasteurization does, but can also decrease the endospore load. Many papers in the literature have shown that relatively low pressures induce endospore germination [35–38]. Thus, during HS-RT, pressure can produce the germination of endospores, and at appropriate pressure levels, their subsequent outgrowth is not fulfilled. As previously discussed, the appropriate pressure level will depend on the characteristics of the juice and the specific microorganisms. In this sense, Pinto et al. [39] reported that storage at 25–100 MPa and room temperature unleashed the germination of *B. subtilis* endospores in previously inoculated carrot juices (pH 6.0). After germination, *B. subtilis* cells grew at 25 MPa and pressures of at least 50 MPa were needed to avoid their proliferation. In contrast, in inoculated apple juice (pH 3.5), a storage pressure of 25 MPa was enough to produce significant reductions in *Alicyclobacillus acidoterrestris* endospores that could not proliferate after germination [40]. In both juices, storage at 50–100 MPa reduced the total microbial load, and the larger the pressure, the quicker the microbial inactivation. For example, Pinto, Martins, Santos, Fidalgo, Delgadillo, and Saraiva [40] reported that *Alicyclobacillus acidoterrestris* spores inoculated in apple juice (10^4–10^5 cells/mL) were inactivated to an undetectable level (<10 CFU/mL) after 30 days at 50 MPa, but after only one day at 100 MPa.

2.2. Enzymatic and Chemical Reactions

It is well-known that pressure affects the rate of enzymatic and chemical reactions. Thus, those reactions with a negative partial activation volume will be enhanced under pressure, while those with a positive partial activation volume will be hindered [41]. Just after fruit juicing, the mixing of fruit enzymes with the substrate and air can rapidly initiate a number of enzymatic reactions that are capable of degrading nutrients and bioactive compounds, modifying pectin, and affecting color or flavor, among others. These enzymatic reactions, together with an endless number of chemical reactions involving oxygen, metal cations, and other juice constituents, can be extended during juice storage and produce significant quality losses [17]. Therefore, to assess the efficacy of hyperbaric storage in preserving fruit juices, it is essential to know which degradative reactions are pressure enhanced and which ones are inhibited.

The effect of pressure on a specific degradative reaction is usually assessed by comparing either the decrease in the reagents implied or the increase in the products formed after some time under different pressure levels. This is not a simple task because several enzymatic and non-enzymatic reactions can be simultaneously involved in the degradation of a specific compound and pressure can affect each of them in a different manner. To avoid the interference of undesired reagents and/or reactions, purified enzymes and food models are frequently used in kinetic studies. However, the results may not represent the real product as environmental factors such as pH and the presence of salts, sugar, or other food constituents can affect the reaction rates under pressure. In contrast, studies in real products frequently fail in identifying the effect of pressure on a specific reaction because many reactions can simultaneously occur, and therefore, only the net effects can be evaluated. Moreover, in this kind of study, microbial metabolism can be an additional factor that interferes with the results, and therefore, actions to prevent microbial growth should be taken.

During hyperbaric storage, enzymatic reactions can be accelerated or decelerated depending on the changes that pressure induces in the structure of the enzyme, the properties of the substrate and/or the solvent (pH, viscosity, density, and so on), or in the reaction mechanisms; for example, a change in the rate-limiting step [42]. Unfortunately, very scarce information exists about the catalytic activity of the enzymes responsible for fruit juice spoilage under pressure. Most studies have focused on pectin methylesterase (PME) or polygalacturonase (PG), two specific enzymes responsible for cloud destabilization and serum viscosity decay, but these are usually performed in model systems and at pressures significantly higher than those employed in HS-RT [43–46]. In general, these studies have shown that the catalytic activity under pressure not only depends on the specific enzyme and the pressure and temperature conditions applied, but also on the enzyme origin, the substrate employed, and the ionic environment.

Recently, Bermejo-Prada et al. [47] evaluated the catalytic activity of strawberry PME at pressure and temperature conditions similar to those employed in HS-RT. To do so, the authors compared the amount of methanol released during the enzymatic reaction at 0.1–200 MPa and 20 °C. They reported that pressure up to 200 MPa did not affect the catalytic activity of PME in the strawberry crude extract at 20 °C. Similar results were found by other authors in crude PME extracts from different plant sources such as tomato [45,48], carrot [44], or pepper [49] at pressures lower than 300 MPa. However, Bermejo-Prada, Segovia-Bravo, Guignon, and Otero [47] noted that unlike in the crude extract, pressure enhanced PME catalytic activity in strawberry juice. Thus, after two days of storage, pectin demethoxylation occurred significantly faster at 200 MPa than at atmospheric pressure or at 50 MPa. To justify these results, the authors suggested that pressure could enhance the activity of some endogenous pectinases, other than PME, that reduced steric hindrances and eased the PME access to methyl ester bonds of pectin. For example, candidates could include pectin- and pectate-lyases as well as debranching enzymes that catalyze changes in pectin side chains such as galactosidases or arabinofuranosidases. In this connection, Bermejo-Prada [50] observed that, at room temperature, pressure up to 200 MPa significantly reduced the catalytic activity of the crude strawberry β-galactosidase extract, and therefore, this enzyme could not be directly implicated in the enhanced pectine demethoxylation observed at 200 MPa. Studies on purified tomato PG also showed a reduced activity of this enzyme under pressure (100–400 MPa) at temperatures between 30 °C and 50 °C [46,51]. However, it is important to note that, as previously commented, results obtained in purified enzymes in buffer solutions may be not representative of real products, therefore more research about the activity of degradative enzymes under pressure is needed.

In real juices, there exists some rough information about the effect of hyperbaric storage on the rate of degradation of some bioactive compounds. To avoid microbial interferences, Bermejo-Prada and Otero [52] added an antimicrobial agent to strawberry juice and observed that HS-RT, either at 50 MPa or at 200 MPa, for 1–14 days did not affect the degradation rate of total phenolics. In contrast, Pinto, Moreira, Fidalgo, Santos, Vidal, Delgadillo, and Saraiva [19] reported that the total phenolics in watermelon juice degraded more quickly at 50–100 MPa than at atmospheric pressure. Differences observed in strawberry and watermelon juices can be due to several reasons such as the composition of the food matrix, pH, or the presence/absence of an antimicrobial agent. Moreover, different enzymatic and non-enzymatic mechanisms could be implied in the results. However, the usual total phenolics analysis in juices by the Folin–Ciocalteau method only gives rough estimations, and more specific methods should be employed to obtain detailed information about the degradation of specific compounds under pressure. In this sense, Bermejo-Prada and Otero [52] showed that HS-RT up to 200 MPa did not affect the net degradation rate of total monomeric anthocyanins (TMA) in strawberry juice, although differences could exist in the mechanisms implied in TMA degradation at atmospheric and high pressure. Thus, significant peroxidase inactivation and lower polymerization were observed in the samples stored at 200 MPa when compared with those maintained at atmospheric pressure.

The stability of carotenoids under pressure has been also explored. Pinto, Moreira, Fidalgo, Santos, Vidal, Delgadillo, and Saraiva [19] analyzed the evolution of lycopene content in watermelon juice during HS-RT and observed that after 10 days at 50 MPa or 100 MPa, the lycopene content was similar to that in conventionally cold stored samples.

3. Effect of Hyperbaric Storage on Juice Quality: Comparison with Conventional Cold Storage

During juice storage, the action of the diverse degradative agents can produce changes in many physical and chemical attributes that finally result in undesirable quality losses in the product. As pressure can inhibit microbial growth, those changes associated with microorganisms can be efficiently avoided during HS-RT, but those associated with enzymatic and chemical reactions can be either enhanced or hampered, depending on the effects of pressure on their specific reaction rates. Furthermore, it is important to note that the rate of degradative reactions depends not only on pressure, but also on other factors such as temperature, pH, the juice composition, or the procedure of juice production, among others, and therefore, general conclusions are not always easy to draw.

Hyperbaric storage at room temperature has been found to be more effective than conventional refrigeration in preserving some physicochemical attributes of fruit juices. Thus, Lemos, Ribeiro, Fidalgo, Delgadillo, and Saraiva [18] observed that the initial pH of watermelon juice was maintained for at least 21 days at 75 MPa and room temperature, while it significantly decreased after seven days at atmospheric pressure and 4 °C. HS-RT at 50–100 MPa was also effective in reducing the significant increase of the titratable acidity observed in watermelon juice after 7–10 days of conventional cold storage [19]. Data in the literature also show that HS-RT can preserve the aroma of fruit juices better than refrigeration. Thus, Bermejo-Prada et al. [53] observed that unlike conventional refrigeration, HS-RT at 50–200 MPa for 15 days did not produce changes in any key aroma compound of strawberry juice. However, other physicochemical attributes are better preserved during conventional cold storage as low temperature can be more effective than pressure in slowing down certain enzymatic and chemical reactions.

During juice storage, pigments can be degraded by both enzymatic and chemical reactions. Moreover, certain enzymes such as PPO, POD, or β-glucosidase are implied in the production of colored compounds that also modify the juice color. Data in the literature show that refrigeration preserves juice color better than HS-RT. For example, Segovia-Bravo, Guignon, Bermejo-Prada, Sanz, and Otero [9] noted lower color changes (ΔE) in strawberry juice conventionally cold stored at 5 °C for 15 days than in juices stored at 25–200 MPa and room temperature. Likewise, Lemos, Ribeiro, Fidalgo, Delgadillo, and Saraiva [18] noted that color changes in watermelon juices kept at 75 MPa and room temperature for 21 days were significantly larger than those observed in juices stored either at 75 MPa and 15 °C, or at atmospheric pressure and 4 °C. However, it is important to note that even though the pressure level applied during HS-RT significantly affects the evolution of lightness, redness, yellowness, hue, or chroma in fruit juices [19,52], color changes during HS-RT are usually very slight. Thus, ΔE values lower than 5 and 2 were detected in watermelon juice after 10 days at 75 MPa [18,19] and 100 MPa [19], respectively, while, in strawberry juice, ΔE values lower than 1.5 were observed after 15 days at 25–220 MPa [9,52]. Although color differences larger than 2–4 are considered perceptible to the naked eye, $\Delta E < 5$ are small in practical terms.

The appearance of a juice is not only affected by the stability of pigments or the formation of colored substances during storage, but also by the juice turbidity or cloudiness. During storage, microbial proliferation can increase juice turbidity, while the decantation of solid particles in suspension decreases juice cloudiness. This decantation is mainly produced by viscosity losses in the juice serum. Serum viscosity affects not only the ability to hold the solid particles of the juice in suspension, but also the mouthfeel. During storage, viscosity usually decreases and this decay is generally attributed to the depolymerization of pectin caused, as previously commented, by the combined action of different endogenous pectinases together with microbial growth that also implies an associated enzymatic activity. Pinto, Moreira, Fidalgo, Santos, Vidal, Delgadillo, and Saraiva [19] and Pinto, Moreira, Fidalgo, Santos, Delgadillo, and Saraiva [20] showed that during storage at atmospheric pressure and room temperature, the cloudiness of watermelon juice increased, probably due to microbial proliferation, and refrigeration was an efficient method to slow down this increase significantly. In contrast, HS-RT at 100 MPa produced a significant decrease in the cloudiness of watermelon juice, most likely due to the pressure-enhanced activity of pectinases that reduced serum viscosity. In this sense, Bermejo-Prada, Segovia-Bravo, Guignon, and Otero [47] observed that at room temperature, the serum viscosity of strawberry juices reduced very quickly and the larger the storage pressure, the greater the viscosity decay. Thus, after only one day of storage, they detected viscosity drops of 42.5%, 55.5%, and 74.5% in strawberry juices kept at 0.1, 50, and 220 MPa, respectively. These viscosity drops are likely to be responsible for cloudiness losses that can be related, in turn, with certain color changes during storage.

4. Juice Stability after Hyperbaric Storage

After hyperbaric storage, fruit juices may not always be immediately consumed or processed, but stored for some time at atmospheric pressure. In these cases, it is important to assess the activity of the surviving microorganisms by comparing microbial loads just after hyperbaric storage and after a certain recovery period at atmospheric pressure as well as the residual activity of the main enzymes implied in fruit juice spoilage.

Even though there is much information about the behavior of the main degradative agents and the quality evolution in fruit juices after pressure treatments, most studies are performed at the conditions usually employed in conventional high-pressure processing [54–56] that, as previously commented, are quite different to those applied in hyperbaric storage. Specific studies about the fruit juice stability after HS-RT are very scarce and the results obtained are commented on in the following paragraphs.

It is well-known that under relatively low pressures such as those usually employed in HS-RT, microorganisms are more likely to be stressed or injured, than killed. Thus, once high pressure is released, cells can repair the injuries and proliferate [21,57]. Data in the literature show that microbial recovery after HS-RT depends on the same factors that affect microbial growth during HS-RT. Thus, the storage pressure plays a significant role and the larger the pressure during HS-RT, the more difficult the microbial recovery is when the product returns to atmospheric conditions. For example, Bermejo-Prada, López-Caballero, and Otero [25] observed that after one day of storage at 25–50 MPa, surviving TAM, LAB, and YM in strawberry juice recovered their cell-proliferating capacity and were able to grow when the juice was maintained at atmospheric pressure and room temperature for three days. In contrast, after one day of storage at 200 MPa, the microorganisms were seriously damaged and counts after three recovery days at atmospheric pressure were under the detection limits (1 CFU/mL for TAM and LAB, and 10 CFU/mL for YM). Moreover, longer storage times make microbial recovery more difficult than shorter ones. Thus, microorganisms in strawberry juices kept at 50 MPa for 1–10 days recovered their cell-proliferating ability after decompression much better than those in juices maintained 15 days at the same pressure level. The product characteristics are also very important for the juice stability after hyperbaric storage. For example, after one day at 100 MPa, surviving TAM, LAB, and YM in strawberry juice hardly grew when the juice was kept at atmospheric pressure and room temperature for three days [25]. In contrast, surviving TAM, LAB, and YM in watermelon juice kept at 100 MPa for 2.5 days were able to grow after pressure release when the juice was stored at atmospheric pressure and 5 °C for 7–14 days [26]. The bacteriostatic effect observed in strawberry juice seems to be due to the low pH of the product as surviving TAM, LAB, and YM were able to form colonies when they were plated on the appropriate media immediately after hyperbaric storage. Thus, low pH in juices could not only enhance microbial damage during pressure storage, but also inhibit the outgrowth of sub-lethally injured cells after decompression. All of the above results clearly show that microorganisms can recover their cell-proliferating capacity after hyperbaric storage, especially after short times at relatively low pressure. Therefore, fruit juices should be appropriately preserved after HS-RT if they are not going to be immediately consumed or processed.

Regarding other degradation agents apart from microorganisms, some data exist on a few enzymes responsible for juice decay. As observed for microorganisms, the residual activity after hyperbaric storage seems to depend not only on the specific enzyme, the pressure applied, or the storage time, but also on the product characteristics.

In general, PME activity significantly decreases during storage, either at atmospheric or at higher pressure, and the longer the storage, the lower the residual PME activity. Thus, Bermejo-Prada, Segovia-Bravo, Guignon, and Otero [47] stored strawberry juice (with an added antibiotic agent) at different pressure levels (0.1, 50, and 200 MPa) for 1–15 days and observed that PME activity decreased more slowly at 200 MPa than at atmospheric pressure or at 50 MPa. The authors attributed this effect to an apparent activation of the enzyme at 200 MPa caused by the pressure-enhanced PME release from small cell wall particles present in the juice. However, after seven days of storage, PME activity was similar in all of the juices stored at different pressures, and after 15 days of storage, the residual PME activities were 56%, 52%, and 57% in juices kept at 0.1, 50, and 200 MPa, respectively. In contrast, Pinto, Moreira, Fidalgo, Santos, Vidal, Delgadillo, and Saraiva [19] observed that in watermelon juice, the higher the storage pressure, the lower the residual PME activity, and after 10 days of storage, they measured residual PME activities of 53.5% and 42.8% in samples stored at 50 MPa and 100 MPa, respectively. These slight differences observed in strawberry and watermelon juices could be due to different factors such as the product characteristics (composition, pH, cell wall particles in the juice, and so on) or the effect of the presence/absence of an antimicrobial agent during storage.

A few data also exist on the residual activity, after HS-RT, of some enzymes responsible for color, aroma, or taste degradation. As observed for PME, the residual activity of peroxidase (POD) decreased after hyperbaric storage. Thus, after 10 days of storage, POD activity reduced to 16.8% in watermelon juices kept at 100 MPa [19], while it was 85% in strawberry juices stored at 200 MPa for 15 days [52]. Polyphenoloxidase (PPO) seems to be less pressure labile than POD. Thus, in strawberry juice, PPO activity increased during storage regardless of the storage pressure, and residual PPO activities of 141%, 159%, and 152% were detected after 15 days at atmospheric pressure, 50 MPa, and 200 MPa, respectively. In contrast, in watermelon juice, PPO activity significantly decreased during storage, but the higher the storage pressure, the larger the residual PPO activity. Thus, after seven days at atmospheric pressure, the residual PPO activity was 9.5%, while it was larger than 50% in the samples stored at 50–100 MPa.

There is very scarce information about the evolution of the quality attributes of fruit juices after hyperbaric storage. Few existing data show that storage under pressure does not accelerate quality losses after expansion if appropriate preservation techniques are subsequently applied. For example, Segovia-Bravo, Guignon, Bermejo-Prada, Sanz, and Otero [9] observed that after 15 days of storage at 25–220 MPa, the color of strawberry juices remained stable for at least 15 days at atmospheric pressure and 5 °C where only a very slight viscosity decay was detected.

5. Consumer Acceptability

Even though the instrumental measurements described in the previous sections provide encouraging evidences of the efficacy of HS-RT in preserving fruit juice quality, it is important to evaluate the consumer perception as instrumental measurements and sensorial analysis are frequently not well correlated.

Hedonic sensory analysis has shown that HS-RT for 15 days at either 100 or 220 MPa did not produce any perceptible change in color, odor, taste, or the overall acceptance of strawberry juice [50]. In contrast, after 15 days at 25 MPa, the hedonic scores of taste and overall acceptance significantly decreased although those of color and odor remained unaltered. This is in agreement with the low color differences ($\Delta E = 1.3$) and the volatile profiles instrumentally measured in these juices. In this sense, after an informal smell evaluation, Lemos, Ribeiro, Fidalgo, Delgadillo, and Saraiva [18] also reported that watermelon juice stored either at 62.5 MPa for 58 days or at 75 MPa for 21 days presented a fresh-like smell, with no sign of off-flavors.

Triangle tests have confirmed that hyperbaric storage at pressures lower than 50 MPa is not able to maintain all of the original organoleptic properties of strawberry juice, therefore, juices stored for 15 days at 25–50 MPa were perceived as different from the juice before storage [58]. Taste changes during hyperbaric storage at 25 MPa were verified, and even though instrumental titratable acidity, pH, and total soluble solids measurements did not change during HS-RT, some judges described the taste of the juices kept at 25 MPa as more sour than that of the juice at day 0 [50]. In contrast, juices stored at 50 MPa were perceived as different more because of their lower viscosity than because of their sour flavor. This was in agreement with the large decrease (≈93%) observed in the instrumental viscosity measurements. When asked about their preferences, most judges preferred the fresh strawberry juice at day 0 than those kept at 25–50 MPa for 15 days.

Cold storage after pasteurization, that is, the conventional storage strategy, was not effective in preserving the original organoleptic properties of strawberry juices either. Surprisingly, after 15 days of storage, the pasteurized juices were preferred to the raw juices at day 0 probably because they were described as sweeter. This sweet flavor could be caused by the caramelization of the sugars contained in the juice during thermal processing. To avoid the influence of caramelization when comparing the consumer preference between conventional cold storage and HS-RT, Bermejo-Prada, Colmant, Otero, and Guignon [58] stored pasteurized strawberry juices either at atmospheric pressure and 5 °C or at 25 MPa and 20 °C for 15 days. Triangle tests in these pasteurized juices revealed that the judges could not notice any sensory difference between them. Therefore, HS-RT at 25 MPa and conventional cold storage preserved their organoleptic properties with the same efficacy.

6. Industrial Implementation

As previously mentioned, many papers in the literature have confirmed the efficacy of HS-RT to extend, in terms of food safety and quality, the shelf-life of not only fruit juices, but also a wide variety of food products. Despite this, HS-RT has not yet been introduced in the industry, probably due to the lack of equipment specially designed for it.

Figure 1 presents a simplified description of how hyperbaric storage could be implemented in a juice factory. Once the juice has been obtained, it would be pumped in tanks and pressurized. Then, the pressurized tanks would be sealed with a valve to avoid pressure release and moved to a warehouse. Here, the pressurized juice would be stored at room temperature until distribution. Depending on the customer demands, distribution could be performed either under pressure or the tanks could be decompressed prior to transport.

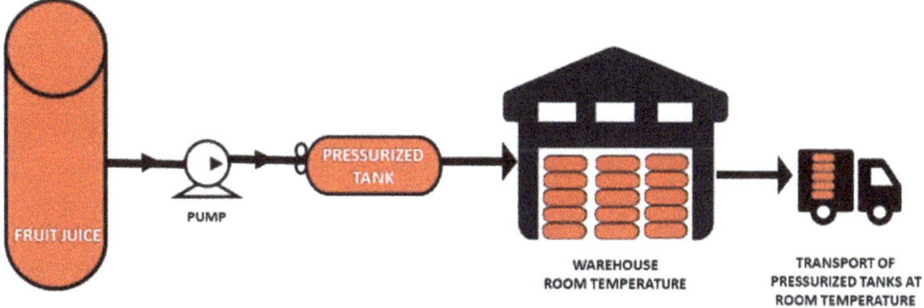

Figure 1. Ideal implementation of hyperbaric storage in a fruit juice factory.

It is clear from Figure 1 that the main components needed in a HS-RT installation are simple: a hydraulic pump and a set of pressure tanks. As pressures employed in HS-RT are quite moderate, the hydraulic pump should not be a problem, but the pressure tanks could present one as currently, tanks do not exist that have been specifically designed for food storage under pressure. Obviously, the feasibility of the logistics management at an industrial scale will depend on the size and weight of these tanks. They must be moved from the pressurization point to the warehouse and then transported to their final destination. Too large or too heavy tanks would make these displacements unviable.

The tank size and weight depend on several factors such as the tank material, the tank shape, the amount of product to store, and the storage pressure (the higher the storage pressure, the thicker the tank wall). Taking into account that typical forklifts can easily move loads up to 3000 kg, Bermejo-Prada, Colmant, Otero, and Guignon [58] deemed pressure tanks larger than 2 m and heavier than 2000 kg (mass of the vessel filled with product) would be difficult to handle inside an industrial facility. They considered cylindrical stainless steel (15-5 PH) tanks, with a fixed diameter/length ratio (0.66), and calculated a domain of viable designs for variable product masses and storage pressures. For a storage pressure of 25 MPa, the largest viable tank should have a length of 1.6 m, a diameter of 1.1 m, and could store 1040 kg of strawberry juice. This volume is much larger than the standard juice batches sold in the industry that usually do not exceed 200 kg. For this mass of juice, the maximal storage pressure for which the tank design would be viable would be 157 MPa. This pressure level is considerably larger than those needed for fruit juice preservation, which are usually between 50 and 100 MPa. Therefore, the construction and handling of pressure tanks appropriate for hyperbaric storage is perfectly viable.

However, logistics management is not the only consideration needed to assess the feasibility of its industrial implementation. The cost of hyperbaric storage is also an important factor to take into account. When calculating the storage cost, the amortization cost of the initial investment, the maintenance cost, and the electricity consumption throughout the storage period must be considered. Taking all of these factors into account, Bermejo-Prada, Colmant, Otero, and Guignon [58] estimated that, today, the HS-RT cost would be considerably higher than that of conventional refrigeration. Even though the electrical consumption in HS-RT is almost negligible, the initial cost of the HS-RT equipment is high and this is, without any doubt, the limiting factor for the implementation of this technology in the food industry. However, it is important to note that the high price of high-pressure equipment has not been an impediment for the successful industrial implantation of conventional high-pressure processing, and thanks to the increasing demand, a decreasing trend in the cost of high-pressure equipment has been observed from 1996 to now. In contrast, the growing tendency of electricity prices could increase the refrigeration cost in the following years and contribute to making this technology less competitive from an economic point of view.

7. Environmental Impact of Hyperbaric Storage

The environmental impact of hyperbaric storage can be roughly estimated by calculating its carbon footprint (CF), that is, the overall emissions of carbon dioxide (CO_2) and other greenhouse gases associated with it. To do so, the equivalent CO_2 emissions associated with both the production of the high-pressure vessel material (direct emissions) and the energy consumption during operation (indirect emissions) must be considered. In this way, Bermejo-Prada, Colmant, Otero, and Guignon [58] estimated that the total CF associated with the storage of 1 kg of strawberry juice at 25 MPa for 15 days was 0.0042 kg $CO_2 \cdot kg^{-1}$ juice. They showed that the HS-RT emissions associated with the vessel material corresponded to almost 100% of the emissions while those derived from the electricity consumption during storage were almost negligible (Figure 2). In contrast, in conventional refrigeration, the total CF was 0.1085 kg $CO_2 \cdot kg^{-1}$ juice, that is, 26 times higher than that of HS-RT and the consumed electricity and refrigerant leakage amounted to almost 95% of the CO_2 total emissions.

Cold facilities are huge consumers of energy and represent about 50% of the total energy consumption in the food industry [5]. Moreover, refrigerants are an important source of GHG emissions and some of them also split and release ozone destructive chlorine atoms. Therefore, HS-RT could represent an important breakthrough for food preservation in terms of refrigerant elimination, energy saving, and environmental protection. Thus, the above results prove that both the needlessness of refrigeration facilities and the extremely low energy requirements (only during compression of the product) represent a real environmental benefit as they contribute to greatly diminish the HS-RT carbon footprint. This is especially important today when global energy savings are demanded in the food industry to fight against climate change.

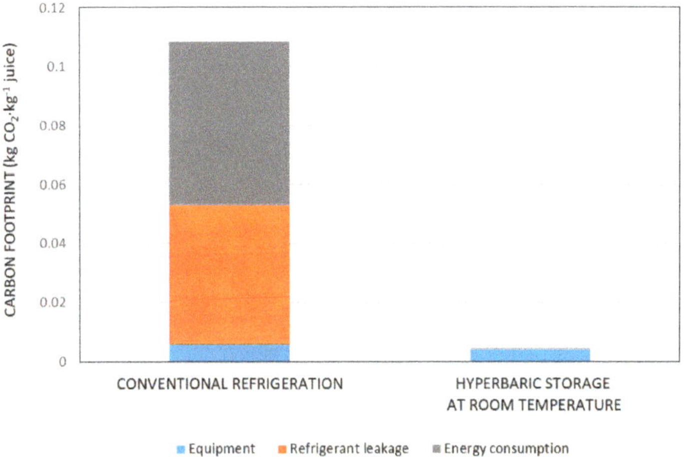

Figure 2. Contribution of different sources to CO_2 emissions in conventional refrigeration and hyperbaric storage at room temperature.

8. Conclusions

In this paper, HS-RT has been presented as an efficient method to preserve fruit juices in terms of juice safety and quality, consumer acceptability, industrial implementation, and environmental impact. Its main advantage over other conventional preservation methods is its extremely low energy consumption. HS-RT could be employed in a wide variety of scenarios: the food industry, ship or truck transport for long distances, school or hospital kitchens, restaurants, or even at home. Moreover, its application in developing countries, where the continuous supply of electric energy is difficult, would be especially noteworthy.

Even though the existing literature analyzed in this paper provides evidence of the feasibility of this method for fruit juice preservation, much more research is still needed. First, HS-RT should be tested on a wide variety of juices obtained from different fruits (e.g., acidic and low-acidic juices, clarified or not, obtained from only one fruit or by mixing a combination of fruit juices and purees) to assess the effect that fruit juice composition can have on quality decay under pressure.

Second, the effect of pressure on the main mechanisms implied in the spoilage of fruit juices should be evaluated in depth. Most existing studies have focused on microorganisms as the main agents responsible for juice spoilage, but the effect of pressure on other degradation agents such as enzymatic and chemical reactions should not be forgotten. In this sense, specific studies to analyze the degradation of relevant compounds (specific vitamins, bioactive compounds, pigments, and so on) under pressure should be designed.

Third, research effort should be particularly focused in identifying issues that could pose a problem for the success of HS-RT such as the adaptation of microbial strains to pressure stress or the pressure-enhanced formation of undesired compounds during storage, among others.

Finally, equipment development for the practical application of HS-RT remains a challenge. Even though pressure tanks do not involve a technological challenge today, their price could be an important limitation. The development of new materials that are able to resist high pressures but are cheaper than steel could help in the industrial success of HS-RT.

Funding: This research received no external funding.

Conflicts of Interest: The author declares no conflict of interest.

References

1. IMARC Group. Fruit Juice Market: Global Industry Trends, Share, Size, Growth, Opportunity and Forecast 2019–2024. Available online: https://www.imarcgroup.com/fruit-juice-manufacturing-plant (accessed on 20 May 2019).
2. AIJN. European Fruit Juice Association. 2018 Liquid Fruit Market Report. Available online: http://viewer.zmags.com/publication/bc62cfea#/bc62cfea/1 (accessed on 20 May 2019).
3. Vermeulen, S.J.; Campbell, B.M.; Ingram, J.S.I. Climate change and food systems. *Annu. Rev. Environ. Resour.* **2012**, *37*, 195–222. [CrossRef]
4. Pelletier, N.; Audsley, E.; Brodt, S.; Garnett, T.; Henriksson, P.; Kendall, A.; Kramer, K.J.; Murphy, D.; Nemecek, T.; Troell, M. Energy intensity of agriculture and food Systems. *Annu. Rev. Environ. Resour.* **2011**, *36*, 223–246. [CrossRef]
5. James, S.J.; James, C. The food cold-chain and climate change. *Food Res. Int.* **2010**, *43*, 1944–1956. [CrossRef]
6. Charm, S.E.; Longmaid, H.E.; Carver, J. Simple system for extending refrigerated, nonfrozen preservation of biological-material using pressure. *Cryobiology* **1977**, *14*, 625–636. [CrossRef]
7. Hirsch, G.P. Method of Pressure Preservation of Food Products. U.S. Patent 5,593,714, 14 January 1997.
8. Hirsch, G.P. Hydraulic Pressure Sterilization and Preservation of Foodstuff and Feedstuff. U.S. Patent 6,033,701, 7 March 2000.
9. Segovia-Bravo, K.A.; Guignon, B.; Bermejo-Prada, A.; Sanz, P.D.; Otero, L. Hyperbaric storage at room temperature for food preservation: A study in strawberry juice. *Innov. Food Sci. Emerg. Technol.* **2012**, *15*, 14–22. [CrossRef]
10. Freitas, P.; Pereira, S.A.; Santos, M.D.; Alves, S.P.; Bessa, R.J.B.; Delgadillo, I.; Saraiva, J.A. Performance of raw bovine meat preservation by hyperbaric storage (quasi energetically costless) compared to refrigeration. *Meat Sci.* **2016**, *121*, 64–72. [CrossRef] [PubMed]
11. Fernandes, P.A.; Moreira, S.A.; Santos, M.D.; Duarte, R.V.; Santos, D.I.; Inacio, R.S.; Alves, S.P.; Bessa, R.J.; Delgadillo, I.; Saraiva, J.A. Hyperbaric storage at variable room temperature-a new preservation methodology for minced meat compared to refrigeration. *J. Sci. Food Agric.* **2019**, *99*, 3276–3282. [CrossRef]
12. Fernandes, P.A.R.; Moreira, S.A.; Duarte, R.; Santos, D.I.; Queirós, R.P.; Fidalgo, L.G.; Santos, M.D.; Delgadillo, I.; Saraiva, J.A. Preservation of sliced cooked ham at 25, 30 and 37 °C under moderated pressure (hyperbaric storage) and comparison with refrigerated storage. *Food Bioprod. Process.* **2015**, *95*, 200–207. [CrossRef]
13. Duarte, R.V.; Moreira, S.A.; Fernandes, P.A.R.; Fidalgo, L.G.; Santos, M.D.; Queirós, R.P.; Santos, D.I.; Delgadillo, I.; Saraiva, J.A. Preservation under pressure (hyperbaric storage) at 25 °C, 30 °C and 37 °C of a highly perishable dairy food and comparison with refrigeration. *CyTA J. Food* **2015**, *13*, 321–328. [CrossRef]
14. Duarte, R.V.; Moreira, S.A.; Fernandes, P.A.R.; Santos, D.I.; Inácio, R.S.; Alves, S.P.; Bessa, R.J.B.; Saraiva, J.A. Whey cheese longer shelf-life achievement at variable uncontrolled room temperature and comparison to refrigeration. *J. Food Process. Preserv.* **2017**, *41*, e13307. [CrossRef]
15. Moreira, S.A.; Fernandes, P.A.; Duarte, R.; Santos, D.I.; Fidalgo, L.G.; Santos, M.D.; Queiros, R.P.; Delgadillo, I.; Saraiva, J.A. A first study comparing preservation of a ready-to-eat soup under pressure (hyperbaric storage) at 25 °C and 30 °C with refrigeration. *Food Sci. Nutr.* **2015**, *3*, 467–474. [CrossRef] [PubMed]
16. Moreira, S.A.; Duarte, R.V.; Fernandes, P.A.R.; Alves, S.P.; Bessa, R.J.; Delgadillo, I.; Saraiva, J.A. Hyperbaric storage preservation at room temperature using an industrial-scale equipment: Case of two commercial ready-to-eat pre-cooked foods. *Innov. Food Sci. Emerg. Technol.* **2015**, *32*, 29–36. [CrossRef]
17. Bates, R.P.; Morris, J.R.; Crandall, P.G. *Principles and Practices of Small-and Medium-Scale Fruit Juice Processing*; FAO Agricultural Services Bulletin; Food and Agriculture Organization of the United Nations: Rome, Italy, 2001; Volume 146.
18. Lemos, Á.T.; Ribeiro, A.C.; Fidalgo, L.G.; Delgadillo, I.; Saraiva, J.A. Extension of raw watermelon juice shelf-life up to 58 days by hyperbaric storage. *Food Chem.* **2017**, *231*, 61–69. [CrossRef] [PubMed]

19. Pinto, C.; Moreira, S.A.; Fidalgo, L.G.; Santos, M.D.; Vidal, M.; Delgadillo, I.; Saraiva, J.A. Impact of different hyperbaric storage conditions on microbial, physicochemical and enzymatic parameters of watermelon juice. *Food Res. Int.* **2017**, *99*, 123–132. [CrossRef] [PubMed]
20. Pinto, C.; Moreira, S.A.; Fidalgo, L.G.; Santos, M.D.; Delgadillo, I.; Saraiva, J.A. Shelf-life extension of watermelon juice preserved by hyperbaric storage at room temperature compared to refrigeration. *LWT Food Sci. Technol.* **2016**, *72*, 78–80. [CrossRef]
21. Abe, F. Exploration of the effects of high hydrostatic pressure on microbial growth, physiology and survival: Perspectives from piezophysiology. *Biosci. Biotechnol. Biochem.* **2007**, *71*, 2347–2357. [CrossRef]
22. Bartlett, D.H. Pressure effects on in vivo microbial processes. *Biochim. Biophys. Acta Protein Struct. Mol. Enzymol.* **2002**, *1595*, 367–381. [CrossRef]
23. Mota, M.J.; Lopes, R.P.; Delgadillo, I.; Saraiva, J.A. Microorganisms under high pressure—Adaptation, growth and biotechnological potential. *Biotechnol. Adv.* **2013**, *31*, 1426–1434. [CrossRef]
24. Aoyama, Y.; Shigeta, Y.; Okazaki, T.; Hagura, Y.; Suzuki, K. Growth inhibition of microorganisms by hydrostatic pressure. *Food Sci. Technol. Res.* **2004**, *10*, 268–272. [CrossRef]
25. Bermejo-Prada, A.; López-Caballero, M.E.; Otero, L. Hyperbaric storage at room temperature: Effect of pressure level and storage time on the natural microbiota of strawberry juice. *Innov. Food Sci. Emerg. Technol.* **2016**, *33*, 154–161. [CrossRef]
26. Fidalgo, L.; Santos, M.; Queirós, R.; Inácio, R.; Mota, M.; Lopes, R.; Gonçalves, M.; Neto, R.; Saraiva, J. Hyperbaric storage at and above room temperature of a highly perishable food. *Food Bioprocess Technol.* **2014**, *7*, 2028–2037. [CrossRef]
27. Lado, B.H.; Yousef, A.E. Alternative food-preservation technologies: Efficacy and mechanisms. *Microbes Infect.* **2002**, *4*, 433–440. [CrossRef]
28. Queirós, R.P.; Santos, M.D.; Fidalgo, L.G.; Mota, M.J.; Lopes, R.P.; Inácio, R.S.; Delgadillo, I.; Saraiva, J.A. Hyperbaric storage of melon juice at and above room temperature and comparison with storage at atmospheric pressure and refrigeration. *Food Chem.* **2014**, *147*, 209–214. [CrossRef]
29. Santos, M.D.; Queirós, R.P.; Fidalgo, L.G.; Inácio, R.S.; Lopes, R.P.; Mota, M.J.; Sousa, S.G.; Delgadillo, I.; Saraiva, J.A. Preservation of a highly perishable food, watermelon juice, at and above room temperature under mild pressure (hyperbaric storage) as an alternative to refrigeration. *LWT Food Sci. Technol.* **2015**, *62*, 901–905. [CrossRef]
30. Matsumura, P.; Keller, D.; Marquis, R. Restricted pH ranges and reduced yields for bacterial growth under pressure. *Microb. Ecol.* **1974**, *1*, 176–189. [CrossRef]
31. Crelier, S.; Robert, M.C.; Claude, J.; Juillerat, M.A. Tomato (*Lycopersicon esculentum*) pectin methylesterase and polygalacturonase behaviors regarding heat- and pressure-induced inactivation. *J. Agric. Food. Chem.* **2001**, *49*, 5566–5575. [CrossRef]
32. Farkas, D.F.; Hoover, D.G. High pressure processing. *J. Food Sci.* **2000**, *65*, 47–64. [CrossRef]
33. Linton, M.; McClements, J.M.J.; Patterson, M.F. Inactivation of *Escherichia coli* O157:H7 in orange juice using a combination of high pressure and mild heat. *J. Food Prot.* **1999**, *62*, 277–279. [CrossRef]
34. Patterson, M.F. Microbiology of pressure-treated foods. *J. Appl. Microbiol.* **2005**, *98*, 1400–1409. [CrossRef]
35. Sarker, M.R.; Akhtar, S.; Torres, J.A.; Paredes-Sabja, D. High hydrostatic pressure-induced inactivation of bacterial spores. *Crit. Rev. Microbiol.* **2015**, *41*, 18–26. [CrossRef]
36. Wuytack, E.Y.; Michiels, C.W. A study on the effects of high pressure and heat on *Bacillus subtilis* spores at low pH. *Int. J. Food Microbiol.* **2001**, *64*, 333–341. [CrossRef]
37. Wuytack, E.Y.; Soons, J.; Poschet, F.; Michiels, C.W. Comparative study of pressure- and nutrient-induced germination of *Bacillus subtilis* spores. *Appl. Environ. Microbiol.* **2000**, *66*, 257–261. [CrossRef]
38. Borch-Pedersen, K.; Mellegård, H.; Reineke, K.; Boysen, P.; Sevenich, R.; Lindbäck, T.; Aspholm, M. Effects of high pressure on *Bacillus licheniformis* germination and inactivation. *Appl. Environ. Microbiol.* **2017**, *83*, e00503–e00517. [CrossRef]
39. Pinto, C.A.; Santos, M.D.; Fidalgo, L.G.; Delgadillo, I.; Saraiva, J.A. Enhanced control of *Bacillus subtilis* endospores development by hyperbaric storage at variable/uncontrolled room temperature compared to refrigeration. *Food Microbiol.* **2018**, *74*, 125–131. [CrossRef]
40. Pinto, C.A.; Martins, A.P.; Santos, M.D.; Fidalgo, L.G.; Delgadillo, I.; Saraiva, J.A. Growth inhibition and inactivation of *Alicyclobacillus acidoterrestris* endospores in apple juice by hyperbaric storage at ambient temperature. *Innov. Food Sci. Emerg. Technol.* **2019**, *52*, 232–236. [CrossRef]

41. Torres, J.A.; Sanz, P.D.; Otero, L.; Pérez Lamela, M.C.; Aranda Saldaña, M.D. Temperature distribution and chemical reactions in foods treated by pressure-assisted thermal processing. In *Processing Effects on Safety and Quality of Foods*; CRC Press: Boca Raton, FL, USA, 2009; pp. 415–440.
42. Eisenmenger, M.J.; Reyes de Corcuera, J.I. High pressure enhancement of enzymes: A review. *Enzyme Microb. Technol.* **2009**, *45*, 331–347. [CrossRef]
43. Duvetter, T.; Fraeye, I.; Sila, D.N.; Verlent, I.; Smout, C.; Clynen, E.; Schoofs, L.; Schols, H.; Hendrickx, M.; Van Loey, A. Effect of temperature and high pressure on the activity and mode of action of fungal pectin methyl esterase. *Biotechnol. Prog.* **2006**, *22*, 1313–1320. [CrossRef]
44. Sila, D.N.; Smout, C.; Satara, Y.; Truong, V.; Van Loey, A.; Hendrickx, M. Combined thermal and high pressure effect on carrot pectinmethylesterase stability and catalytic activity. *J. Food Eng.* **2007**, *78*, 755–764. [CrossRef]
45. Van Den Broeck, I.; Ludikhuyze, L.R.; Van Loey, A.M.; Hendrickx, M.E. Effect of temperature and/or pressure on tomato pectinesterase activity. *J. Agric. Food Chem.* **2000**, *48*, 551–558. [CrossRef]
46. Verlent, I.; Van Loey, A.; Smout, C.; Duvetter, T.; Hendrickx, M.E. Purified tomato polygalacturonase activity during thermal and high-pressure treatment. *Biotechnol. Bioeng.* **2004**, *86*, 63–71. [CrossRef]
47. Bermejo-Prada, A.; Segovia-Bravo, K.A.; Guignon, B.; Otero, L. Effect of hyperbaric storage at room temperature on pectin methylesterase activity and serum viscosity of strawberry juice. *Innov. Food Sci. Emerg. Technol.* **2015**, *30*, 170–176. [CrossRef]
48. Verlent, I.; Loey, A.V.; Smout, C.; Duvetter, T.; Nguyen, B.L.; Hendrickx, M.E. Changes in purified tomato pectinmethylesterase activity during thermal and high pressure treatment. *J. Sci. Food Agric.* **2004**, *84*, 1839–1847. [CrossRef]
49. Castro, S.M.; Loey, A.V.; Saraiva, J.A.; Smout, C.; Hendrickx, M. Identification of pressure/temperature combinations for optimal pepper (*Capsicum annuum*) pectin methylesterase activity. *Enzyme Microb. Technol.* **2006**, *38*, 831–838. [CrossRef]
50. Bermejo-Prada, A. Hyperbric Storage of Foods at Room Temperature: Characterization in Strawberry Juice. Ph.D. Thesis, Complutense University of Madrid, Madrid, Spain, 2015.
51. Verlent, I.; Smout, C.; Duvetter, T.; Hendrickx, M.E.; Van Loey, A. Effect of temperature and pressure on the activity of purified tomato polygalacturonase in the presence of pectins with different patterns of methyl esterification. *Innov. Food Sci. Emerg. Technol.* **2005**, *6*, 293–303. [CrossRef]
52. Bermejo-Prada, A.; Otero, L. Effect of hyperbaric storage at room temperature on color degradation of strawberry juice. *J. Food Eng.* **2016**, *169*, 141–148. [CrossRef]
53. Bermejo-Prada, A.; Vega, E.; Pérez-Mateos, M.; Otero, L. Effect of hyperbaric storage at room temperature on the volatile profile of strawberry juice. *LWT Food Sci. Technol.* **2015**, *62*, 906–914. [CrossRef]
54. Gupta, R.; Balasubramaniam, V.M. High-pressure processing of fluid foods. In *Novel Thermal and Non-Thermal Technologies for Fluid Foods*; Cullen, P.J., Brijesh, K.T., Valdramidis, V.P., Eds.; Academic Press: San Diego, CA, USA, 2012; pp. 109–133.
55. Koutchma, T.; Popović, V.; Ros-Polski, V.; Popielarz, A. Effects of ultraviolet light and high-pressure processing on quality and health-related constituents of fresh juice products. *Compr. Rev. Food Sci. Food Saf.* **2016**, *15*, 844–867. [CrossRef]
56. Chakraborty, S.; Kaushik, N.; Rao, P.S.; Mishra, H.N. High-pressure inactivation of enzymes: A review on its recent applications on fruit purees and juices. *Compr. Rev. Food Sci. Food Saf.* **2014**, *13*, 578–596. [CrossRef]
57. Huang, H.-W.; Lung, H.-M.; Yang, B.B.; Wang, C.-Y. Responses of microorganisms to high hydrostatic pressure processing. *Food Control* **2014**, *40*, 250–259. [CrossRef]
58. Bermejo-Prada, A.; Colmant, A.; Otero, L.; Guignon, B. Industrial viability of the hyperbaric method to store perishable foods at room temperature. *J. Food Eng.* **2017**, *193*, 76–85. [CrossRef]

Review

Potential Applications of High Pressure Homogenization in Winemaking: A Review

Piergiorgio Comuzzo * and Sonia Calligaris

Department of Agricultural, Food, Environmental and Animal Sciences, University of Udine, via Sondrio 2/A, 33100 Udine, Italy

* Correspondence: piergiorgio.comuzzo@uniud.it; Tel.: +39-0432-55-8166

Received: 17 June 2019; Accepted: 26 August 2019; Published: 3 September 2019

Abstract: High pressure homogenization (HPH) is an emerging technology with several possible applications in the food sector, such as nanoemulsion preparation, microbial and enzymatic inactivation, cell disruption for the extraction of intracellular components, as well as modification of food biopolymer structures to steer their functionalities. All these effects are attributable to the intense mechanical stresses, such as cavitation and shear forces, suffered by the product during the passage through the homogenization valve. The exploitation of the disruptive forces delivered during HPH was also recently proposed for winemaking applications. In this review, after a general description of HPH and its main applications in food processing, the survey is extended to the use of this technology for the production of wine and fermented beverages, particularly focusing on the effects of HPH on the inactivation of wine microorganisms and the induction of yeast autolysis. Further enological applications of HPH technology, such as its use for the production of inactive dry yeast preparations, are also discussed.

Keywords: high pressure homogenization (HPH); wine technology; microbial inactivation; ageing on lees; yeast autolysis

---

1. Introduction

In the last two decades, there has been an increasing interest of the winemaking sector towards innovative technologies, particularly certain physical, non-thermal processing methods, able to improve process efficiency and reduce the use of chemical inputs. Ultrasounds (US), pulsed electric fields (PEF) or high hydrostatic pressures (HHP) are able to increase the mass transfer rate from grape skins [1], allowing polyphenols extraction and shortening certain wine production steps (e.g., skin maceration) [2]. In addition, they may decrease must and wine microbial populations, facilitating the reduction of sulfur dioxide use [3], an important goal of modern enology, due to the toxicity and allergenic potential of this additive [4]. Currently, in Europe, the use of these technologies are still not allowed at winery scale [5]. However, some of them, e.g., US or PEF, are in the course of evaluation by the International Organization of Vine and Wine (OIV) and presumably, they will be included soon in the OIV International Code of Oenological Practices [6].

Among these novel technologies, high pressure homogenization (HPH) is the least studied and only a limited number of papers have reported scientific results about its use for must and wine processing. Outside winemaking, HPH is utilized in several areas, such as chemical, pharmaceutical, biotechnology, and food industries. Emulsion, dispersion, encapsulation, and mixing are among the traditional reasons for employing homogenization operation during fluid food processing. However, beside these traditional functions, HPH is nowadays claimed to be one of the most promising novel non-thermal technologies that can be applied to improve fluid food safety and quality. It is a matter of fact that different potential applications of HPH have emerged in the last decades from both industrial and scientific efforts. For this reason, considering the strong leaning towards innovation of modern enology, HPH may represent an interesting perspective for winemaking applications. Today several HPH devices are available for lab scale experiments but also ready for the scaling up at an industrial level to process and design novel foods with improved functionalities.

Based on these considerations, this review will critically discuss the possible applications of HPH in the wine sector, highlighting relevant results reported in literature, advantages as well as possible drawbacks of this technology, particularly considering its application in a rationale of reduction of chemical input and maximization of the quality of wine production.

2. Basic Principles of High Pressure Homogenization and Process Design

The term "homogenization" refers to the physical process allowing the reduction of the particle size of a polydisperse liquid system. The main result is an increased number of smaller particles of a narrow size range [7]. The most common installations used to this purpose are today based on valve technology: A fluid is forced to pass through the homogenization valve and the mechanical forces suffered by the product lead the disaggregation of particles. Therefore, the total surface area of newly formed particles increases, leading to a significant improvement of product physical stability. Standard homogenization at pressure levels around 20–50 MPa is widely used in chemical, pharmaceutical, biotechnology, and food industries, mainly to reduce the particle size of liquid dispersions, preventing phase separation during storage.

The current development of homogenizers have allowed much higher pressures leading to the development of high pressure homogenization (HPH) technology. This technology is based on the same principle of conventional homogenization, but it works at significantly higher pressures, up to 400 MPa. Depending on the nominal pressure level, the technology is called high pressure homogenization (HPH, up to 150–200 MPa) or ultra-high pressure homogenization (UHPH, up to 350–400 MPa) [8]. Basically, during the HPH process, the fluid is forced to pass through a narrow gap in the homogenizer valve, where it is submitted to a rapid acceleration [9]. Therefore, the suspended materials in the fluid are subjected to great mechanical forces and elongation stresses, becoming twisted, deformed, and disrupted [9].

The design of the homogenization valve and its geometry is of a special importance to generate the desired effect on processed fluid. Homogenizers are usually equipped with a high pressure homogenization valve that must resist to the intense mechanical stresses generated during the fluid passage and a low-pressure valve allowing stresses reduction before fluid exit (Figure 1). Nowadays, there is a number of commercially available valves with different geometries and made of different materials, as extensively described by Martinez-Monteagudo et al. [10]. Basically, these authors classified HPH valves into counter-jet, radial diffusers, and axial flow valves.

It should not be underestimated that independently on valve geometry, the liquid travelling through the HPH valve is accompanied by intense heating phenomena, causing the liquid temperature increase. The fluid temperature increases linearly with the pressure by 14–18 °C per 100 MPa due to shear effects and the conversion of mechanical forces into heat [8]. The temperature changes during HPH should be carefully taken into account during process design to avoid possible undesired thermal damages to the product or, contrarily, to exploit temperature as additional factor, beside pressure, to generate the desired changes to the product.

The treatment intensity and relevant effects can be modulated by applying different operative pressures and number of passes at the selected pressure. To quantify the process intensity, the energy density (E_v, MJ/m^3) transferred from the homogenization valve to the sample can be determined as described by Stang et al. [11] and Calligaris et al. [12]:

$$E_v = \Delta P \cdot n \quad (1)$$

where ΔP is the pressure difference operating at the nozzles and n the number of passes at the selected pressure. Basically, process intensity is directly related to the homogenization pressure and number of consecutive passes through the valve.

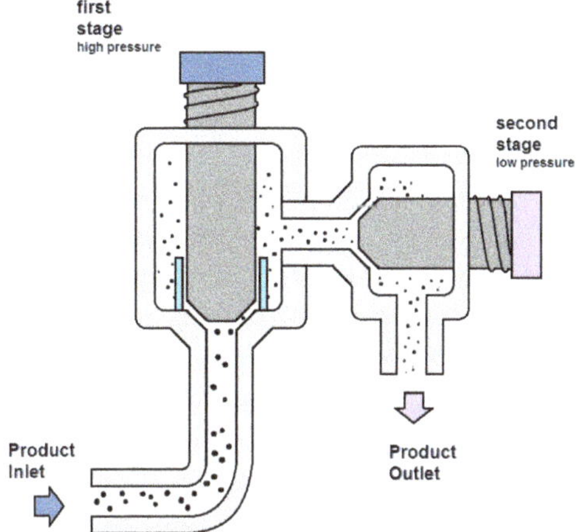

Figure 1. Scheme of a two-stage high pressure homogenizer. Modified From Plazzotta and Manzocco [13].

3. Possible Applications of HPH in the Food Sector

The intense mechanical stresses suffered by the product during the HPH process can induce different modifications in food constituents. The most widely studied effect of this technology is the cell disruption capacity to induce microbial inactivation and facilitate the removal of intracellular components. However, more recently, HPH has been proposed as an efficient tool to steer the structure and functionality of food biopolymers. For this reason, different food sectors are interested in understanding the feasibility of this technology for specific applications. Table 1 shows the main possible uses of HPH technology in the food sector and relevant recent reviews on this topic. Beside these specific papers, other general reviews have also been published on HPH and UHPH [8,10,14,15]. Below, a brief summary of the main possible application of HPH on fluid foods and beverages is reported.

Table 1. Recent reviews on specific applications of high pressure homogenization.

Application	Literature Reviews
Emulsification—Nanoemulsification	[16–18]
Microbial inactivation	[19,20]
Cell disruption and recovery of intracellular components	[21,22]
Physical and physical–chemical modifications of food biopolymers	[23]
Enzyme inactivation	[24]

3.1. *Emulsification and Nanoemulsification*

The droplet size reduction generated by the application of a homogenization process is of key importance in the production of stable emulsion and, thus, to obtain the desired quality of the final product. The possibility of increasing the pressure level from conventional homogenization to higher pressures has favored the application of this technology to produce nanoemulsions that are heterogeneous systems containing particles with droplets diameter lower than 200 nm [16,17]. During a single pass process, a progressive size reduction can be obtained by increasing the homogenization pressure. However, up to a certain pressure level, which depends on the plant design and emulsion formulation, particle size reduction is no longer expected [8,12,25–27]. Multiple passes through the homogenization valve are eventually applied to further reduce not only the mean particle diameter but also the width of the particle size distribution, improving emulsion stability against coalescence [12,25,26].

3.2. Microbial Inactivation

The majority of the studies deals with the possibility of using HPH to disrupt microbial cells and thus as an alternative technology to conventional heat-treatments [8,19,28]. Good results have been obtained for HPH inactivation of microorganisms and foodborne pathogens present in different food substrates. The capacity of HPH to induce microbial inactivation has been attributed to the high pressure, velocity gradients, shear stresses, turbulence, shocks, and cavitation phenomena occurring during the passage through the homogenization valve. These stresses induce cell membrane permeabilization, followed by the deformation of the cell structure and cytoplasmatic organelles, and leaking of intracellular materials. Moreover, the temperature increase suffered by the fluid during the process could improve the microbial inactivation efficacy of HPH [19]. Indeed, additional or synergy effects between mechanical forces and heat have been described by different authors at temperatures higher than 60 °C [28–30].

In general, gram-positive bacteria show a higher resistance to HPH than gram-negative ones [31,32]. This effect was associated to the thinner cell walls membrane of gram-negative bacteria in comparison to gram-positive. Yeast and fungi exhibited resistance to HPH, intermediate between gram-negative and gram-positive bacteria. In fact, even if they have a thicker membrane than gram-positive bacteria, their larger size as well as different cell wall composition, rich in glucans, mannans and proteins, reduced their resistance to the HPH stresses [33,34]. Finally, spores are the most resistant to high pressure homogenization and their inactivation by HPH or UHPH is still a challenge. Literature data have pointed out that spore inactivation can be obtained only by defining a proper combination of pressure, temperature, and additional chemical hurdles [35–37].

It can be concluded from literature that HPH can be reliably applied for microbial inactivation, and has been proposed as alternative to food pasteurization or, in some cases, sterilization [15]. Promising results have been obtained with cow milk [30], vegetable milks [38], and fruit juices [28,39,40]. Overall, the reduction of microbial count is pressure dependent, showing better inactivation as homogenization pressure increased. However, other process parameters such as pressure level, inlet temperature of processed fluid, number of recycling passes, should be accurately defined, taking into account not only the target microbial strain and level of inactivation but also food characteristics in term of chemical composition and physical properties.

3.3. Cell Disruption and Recovery of Intracellular Components

Mechanical cell disruption by HPH is routinely exploited in pharmaceutical and biotechnological industries to disrupt bacteria and yeast in the attempt to recovery bioproducts [22]. Today it is claimed to be an efficient tool for cell disruption also in the food sector and it can be potentially exploited as pre-treatment to facilitate the recovery of valuable components from vegetable wasted material [13,41,42] or from microalgae [21,43]. The cell disruption degree is reported to be highly dependent on matrix characteristics and HPH intensity in term of operating pressure and number of passes through the homogenization valve [44,45].

3.4. Physical and Physical-Chemical Modifications of Food Biopolymers

Beside the direct effect of HPH on cells, a wide number of papers deal with the possibility of applying HPH treatments to modify the physical and physical–chemical characteristics of food components, mainly proteins and polysaccharides.

Regarding proteins, it has been reported that protein structure can be modified during the intense mechanical stresses suffered by the product passing through the homogenization valve. Because of protein conformational changes induced by HPH, protein functionalities, in terms of solubility, emulsifying, gelling, or foaming capacity, could deeply change. HPH treatments were able to affect the physicochemical properties of a number of proteins including milk proteins [46], soybean proteins [47,48], myofibrillar proteins [49], peanut proteins [41], mussel proteins [50], and egg white proteins [51].

Similarly, HPH can induce severe changes in polysaccharides dispersions, such as mass, hydrodynamic radius, and viscosity. On this regards, studies on carboxymethlycellulose, guar gum, hydroxymethylcellulose [52], starch [53], methylcellulose [54], inulin [55], and pectin [44] are available in literature.

3.5. *Enzyme Inactivation*

Since as previously reported, HPH can modify protein structure, this technology has been proposed as tool to reduce the activity of indigenous enzymes. According to literature, HPH can increase or decrease enzyme activity depending on the enzyme source, processing conditions as well as matrix characteristics [24]. The modification of enzyme activity has been attributed to the changes of protein native structure associated to the mechanical forces and temperature increase during HPH process. In fact, protein unfolding and dissociation of native oligomers can be obtained because of pressurization, leading to changes of enzyme activity [48]. In agreement with literature, it can be said that complete enzyme inactivation can be difficult to reach by single pass homogenization, even at the highest homogenization pressures during UHPH. The complete inactivation can be achieved only by increasing temperature and/or the number of recycling passes [24,56].

4. Potential Use of HPH in Winemaking

The first attempts of using high pressure technologies for must and wine processing date back to the 1990s. Hyperbaric treatment initially consisted of applying high pressure processing in hydrostatic conditions [57]. Nowadays, the effects of high hydrostatic pressure on the extraction of color and phenolic compounds from grapes [1,2,58,59], as well as on the inactivation of wild microorganisms in grape, juice, and wine [2,3,57,59–61], are well known.

Nevertheless, compared with high hydrostatic pressure (HHP), high pressure homogenization (HPH) is based on different principles. In fact, HHP is a static batch processing technique, in which the products are pre-packaged and introduced into a pressurized chamber for a given time [62]. In contrast, HPH is a dynamic high pressure technology and the modifications it induces on plant tissues and microbial cells do not depend on the application of pressurization alone, since they are also affected by other physical phenomena such as cavitation, turbulence, and shear, into the homogenizing valve (Sections 2 and 3). For this reason, HPH is not suitable for the processing of grape mash after crushing–destemming, because of the presence of skins and seeds, which may clog the valves of the homogenizer. However, compared to HHP, it allows continuous in-flow processing and, after a preliminary preparation of the fluid (must or wine) by the elimination of skin fragments, seeds, crystals, and other solid particles, it may be more suitable for the treatment of musts and wines in the volumes normally found on a winery scale.

In the following paragraphs, the most interesting applications of HPH technology for winemaking use will be discussed. They are the inactivation of spoilage microorganisms in grape juice and wine, the acceleration of the yeast autolytic process and the production of yeast derivative preparations to be used as processing aids.

4.1. *Control of Microbial Populations in Grape Juice and Wine*

The effect of high (hydrostatic) pressures on wine microbial populations is well known and it has been reported since 1995 [57]; thirteen wine yeasts species including *Lactobacillus* spp., *Oenococcus oeni*, *Acetobacter* spp., and *Botrytis cinerea*, added at 10^6 CFU/mL to a Moscato wine, were inactivated in 2 min by a HHP treatment at 400 MPa. The positive effect of the application of high pressure in static conditions on the inactivation of wine microorganisms was further confirmed in the following years [63]. Mok and co-workers [61] found that total yeasts (2.9 × 10^5 CFU/mL) were completely eliminated from red wine in 30 min at 300 MPa and in 10 min at 350 MPa; at the latter pressure value, lactic acid bacteria (LAB, 2.9 × 10^5 CFU/mL) were destroyed in 5 min. More recently, Morata and colleagues [2] reported significant reduction of yeasts (up to 4 Log units), total aerobic bacteria, and LAB (up to approx. 1 Log unit) in crushed Tempranillo grapes treated by HHP (200, 400, and 550 MPa).

Despite these positive results and the advantage (compared with HHP) of the possibility of using homogenization for continuous in-flow processing, HPH has been poorly exploited for must and wine treatment. Recently, Loira et al. [64] tested UHPH on white must, in comparison with sulfiting (SO_2 35 mg/L) and untreated juice. UHPH (300 MPa) determined the complete elimination of wild yeasts (initial load 1 × 10^6 CFU/mL) from the treated samples, while the microbial load of the sulfited must was not significantly changed with respect to the untreated control juice. In addition, UHPH-processed must, stored at 18 °C without yeast inoculation, showed the absence of fermentation for eight days. The same trend observed for yeast was also found for wild bacteria (LAB and aerobic bacteria): UHPH treatment (300 MPa) reduced below the limit of detection (LOD 1 CFU/mL) the initial load of 7 × 10^3 CFU/mL [64]. It is interesting to observe that the extent of heating connected with UHPH treatment reported in this study is extremely low (inlet homogenizer temperature: 20 °C; outlet temperature: 25 °C).

Puig et al. [65] tested HPH for reducing indigenous flora in Parellada and Trepat musts. Results showed that HPH treatment (200 MPa) was able to completely eliminate LAB (3 and 5 Log units reduction, respectively, in Parellada and Trepat), fungi and yeasts (3 and 6 Log units reduction, respectively, in Parellada and Trepat) in both musts, and only a limited residual population of total bacteria (other than LAB) were detected.

Using lower pressure values, Comuzzo and colleagues [66] found less evident results applying HPH on *Saccharomyces bayanus*; an active dry yeast preparation (ADY) was rehydrated, processed by 1–10 passes at 150 MPa and freeze-dried. HPH was carried out in conditions of uncontrolled and controlled temperature regimes, by positioning a heat exchanger at the homogenizer outlet. Results showed that the initial load of the ADY preparation (approximately 10 Log CFU/g) decreased as the number of passes increased, but if the temperature was controlled (T_{out}, at homogenizer outlet 32 °C, in conditions of the controlled temperature regime), the maximum decrease of total yeast population was lower than 4 Log CFU/mL. In contrast, a temperature increase, in the conditions of uncontrolled heating (without heat exchanger), gave an important contribution: Yeast viability was reduced at 1.9 Log units after the 6th pass (T_{out} 70 °C) and below LOD (10 CFU/mL) after the 10th pass (T_{out} 74 °C).

The intensity of the pressure applied may also influence the efficiency of HPH in eliminating wine yeasts. The same authors [67] reported that the inactivation of a rehydrated *S. bayanus* commercial active yeast strain increased linearly by increasing the pressure, despite that the maximum pressure applied (150 MPa) provoked only a diminution of approximately 2 Log units (Table 2).

Table 2. Application of HPH (50–150 MPa) to a water suspension of *S. bayanus* active dry yeast (ADY); effect on yeast viability. Extracted from Comuzzo et al., 2015—Modified [67].

Sample	Total Yeasts (Log CFU/g)
ADY [1]	10.8
50 MPa [2]	9.9
100 MPa [2]	9.2
150 MPa [2]	8.6

[1] ADY: Active Dry Yeast; [2] analyzed after HPH and freeze-drying.

The effects of HPH on microorganisms of enological interest are evident also in other papers related to fruit juices or other fermented beverages: *S. bayanus* inactivation in apple juice [37]; *S. cerevisiae* and *Lactobacillus plantarum* in orange juice [68]; *S. cerevisiae* and *Lactobacillus delbrueckii* in orange, apple, and pineapple juice [40]; different strains of *Lactobacillus* spp., *Pediococcus* spp., *Acetobacter aceti*, and *S. ludwigii* in beer [69]; and total yeasts in rice wine [70].

Generally speaking, pressures higher than 250 MPa allow a complete inactivation of microorganisms [68]; contrarily, lower pressure values generally require multi-pass processing [40,69], or the combination of mechanical forces and heating [69].

4.2. Acceleration of Yeast Autolysis and Ageing on Lees

Autolysis is the self-degradation of yeast cell constituents that begins after cell death, promoted by the lytic activity of cellular enzymes [71]. Autolytic phenomena are fundamental during ageing on the lees (*élevage sur lies*), an important technological tool for the production of certain wine typologies, such as white wines from Burgundy or sparkling wines produced by the traditional method (e.g., Champagne, Cava or Franciacorta).

Despite different papers highlighted the possibility of using homogenization for promoting the extraction of intracellular components from *Saccharomyces* cells [22,72–74], very few publications report original data concerning the ability of HPH to accelerate yeast autolysis in wine-like media.

The first evidence about the possibility to use HPH for this purpose was reported by Patrignani et al. [75]. The authors applied HPH treatment (90 MPa) on different yeast strains (*S. cerevisiae* and *S. bayanus*) prior to their use for the preparation of *tirage* solutions, for sparkling wine refermentation (traditional method). The treatment poorly affected yeast viability and refermentation behavior (all the strains allowed to reach a final overpressure of approximately 6 bars), but scanning electron microscopy highlighted that HPH provoked an acceleration of autolysis over a 40 day ageing period. The authors hypothesized that HPH might presumably activate the enzymatic pool involved in autolytic process; it is interesting to observe that temperature was controlled during the experiment: Inlet temperature was 25 °C and the samples were immediately cooled to 3 °C after the treatment. In such a way, the influence of temperature can be excluded.

The effect of pressure and number of passes on the extraction of intracellular components from *Saccharomyces* cells was described also in other publications [22,76]: The release of ionic compounds, proteins, and other bio-active compounds significantly increased by increasing these two operating parameters.

The possibility to accelerate yeast autolysis is an interesting perspective in winemaking practice. In fact, it is well known that prolonged ageing on the lees may increase the risk of microbial spoilage and production of unwanted metabolites, such as biogenic amines [77]. This risk might be further reduced by HPH treatment, due also to the ability of such technology to reduce wild microorganisms and LAB (Section 4.1).

An interesting approach for managing ageing on lees through HPH technology was described by Carrano [78]; fresh lees were treated at 60 and 150 MPa (single pass) and added to a white wine for ageing on lees. HPH increased the ability of the treated lees to release glucidic colloids in model wine, also determining a significant reduction of viable yeasts and LAB (Table 3). This approach may potentially allow the reduction of the use of sulfur dioxide during ageing on lees, when HPH-processed lees are re-incorporated into the wine.

Table 3. Application of HPH (60 and 150 MPa) on fresh lees; effect on yeast and LAB populations. Extracted from Carrano, 2016—Modified [78].

Sample	*Saccharomyces* spp. (Log CFU/mL)	NON *Saccharomyces* spp. (Log CFU/mL)	LAB (Log CFU/mL)	Temperature (°C) [2]
Untreated	1.9	1.3	3.9	-
60 MPa	1.4	0.7	3.1	40
150 MPa	*n.d.* [1]	*n.d.*	*n.d.*	50

[1] *n.d.*: not detected (<10 CFU/mL); [2] measured at homogenizer outlet.

4.3. Production of Yeast Derivative Preparations for Enological Use

Yeast derivatives (YDs, inactive dry yeasts, and yeast autolysates) are processing aids largely used in the wineries for several purposes, such as nutrients for yeasts or LAB starter cultures or colloidal supplements during wine ageing [79]. They are produced from *Saccharomyces* spp. by natural autolysis (i.e., through the action of endogenous enzymes), combined with heat treatment and/or modification of the pH [80]. Given their mode of preparation, the production of YDs may be considered as a special case of management of the autolysis process; for this reason, HPH might also be exploited for the industrial manufacture of these products.

Comuzzo and colleagues [66,67] studied the potential use of HPH for the production of yeast autolysates for winemaking. HPH was able to increase the ability of an ADY preparation of *S. bayanus* to release glucidic colloids, proteins, and amino acids in wine-like solution. This effect was proportional to the pressure applied (0–150 MPa) [67] (Table 4).

Table 4. Application of HPH (50–150 MPa) to a water suspension of *S. bayanus* active dry yeast (ADY); effect on the release of proteins and glucidic colloids in wine-like solution. Extracted from Comuzzo et al., 2015—Modified [67].

Code	Soluble Proteins [2] (mg/g)	Total Colloids [2] (mg/g)
ADY [1]	14	43
50 MPa	36	101
100 MPa	47	128
150 MPa	51	165

[1] ADY: Active Dry Yeast; [2] Amount released in wine-like solution (pH 3.2, ethanol 12% *v*/*v*) by one gram of ADY or HPH-processed yeast after freeze-drying.

Moreover, the number of passes (1–10) and processing temperature affected the composition of the autolysates and the release of soluble compounds in model wine [66]: lower temperatures, led to higher concentrations of soluble amino acids and proteins, while heating (in conditions of uncontrolled temperature regime—T_{out} 74 °C) provoked a decrease of the amounts of these two groups of molecules. The authors have suggested that HPH processing conditions might be differently set up for tailoring the characteristics of the autolysate, making them suitable for different applications during wine production. For instance, for obtaining a nitrogen supplement for alcoholic fermentation, processing temperature must be controlled to maximize the content of free amino acids useful as nutrient for yeast growth. In contrast, if a colloidal supplement is needed during wine ageing (e.g., for improving wine mouthfeel characteristics), processing temperature could be kept higher, with the advantage of reducing amino acid content and improving microbiological stability during wine storage and ageing [66].

Finally, concerning the possible advantage of using HPH for the production of YDs, it is well known that such kind of products may sometimes negatively affect wine volatile composition, due to the release of exogenous aroma compounds in wine [81,82]. The aroma composition of HPH-processed yeast autolysates differs to that of the products obtained by conventional thermolysis methods, because of a higher concentration of ethyl esters and lower amounts of short-chain fatty acids and carbonyl compounds [67]. The latter are probably connected with lipid fraction and its oxidation, and previous experiments have highlighted that they can be released into the wine, affecting its sensory characters [81].

4.4. Modifications Induced on Wine

Few data are available about the modifications induced by high pressures on wine characteristics. Such modifications are mainly connected with the thermal stress suffered by the product during hyperbaric treatment. For this reason, high pressure treatments, both static processing (HHP) and homogenization (HPH), were mainly tested on grapes [2,60], musts [57,64,65], grape by-products [1], or lees [78].

The opportunity to control processing temperature by placing a heat exchanger at the homogenizer outlet might reduce the thermal damage to the product. Keeping temperature below 25 °C, the few papers available report that pressurization had a minor impact on wine color and sensory characteristics, both in static [61] and homogenization conditions [65]. At the same maximum processing temperature, Loira et al. [64] found that the white wines obtained from the fermentation of UHPH-processed musts were more fruity and with better aroma than the control (obtained by spontaneous fermentation) and sulfited samples (inoculated with the same *S. cerevisiae* strain as UHPH-treated must). However, UHPH processing of must led to wines with a higher color intensity in comparison with sulfiting (35 mg/L SO_2).

Apart from the effects of temperature, pressurization alone may also influence wine characteristics. Santos and co-workers [83] analyzed the composition of HHP-processed red wines (500 MPa, 5 min, 20 °C), in comparison with a control (stored in stainless steel vats) and with the same wine aged in oak barrels, in stainless steel vats with oak chips, and in stainless steel vats with oak chips plus microoxygenation. After a storage period of 5 months, pressurized wines showed a lower content of monomeric anthocyanins, phenolic acids, and flavonols in comparison with the other wine treatments; in contrast, HHP promoted wine evolution leading to a higher degree of tannin polymerization and pyranoanthocyanins concentration, similar to those found in the samples obtained with microoxygenation and wood contact.

Talcott and co-workers investigated the effects of thermal treatment and HHP (600 MPa for 15 min) on the color and phytochemical stability of Muscadine grape juice, in combination with ascorbic acid and rosemary extract. HHP determined a slight loss of juice color and antioxidant activity with respect to control and thermally-treated samples, and this effect was proportional to the concentrations of added ascorbic acid and rosemary extract [84]. The authors report that the greater loss obtained after HHP processing was likely due to residual activity of polyphenoloxidase enzymes.

Other papers have investigated the effects of high pressure processing (HHP) in comparison with sulfiting and a control wine produced with no preservation treatment (SO_2-free) [85,86]. Pressurized wines developed a more brownish color and a slightly lower antioxidant activity during one year of storage [86]. In addition, high pressure processing determined a decreased content of free amino acids and a higher concentration of volatile furans, ketones, and aldehydes [85,86], symptoms of a greater extent of Maillard reaction in HHP processed wines.

Based on the evidence, it is clear that most of the data available about the effects of high pressure treatments on wine composition is related to HHP, while, to the best of the authors' knowledge, very few scientific publications have regarded HPH. Although the possibility of controlling processing temperature may allow the application of HPH to the wines, minimizing thermal damage, there are still too few data concerning the modifications induced by the treatment on the compositional and sensory characteristics of wine itself. For this reason, further experiments need to be carried out, considering a major number of processing conditions, analytical parameters, and wine varieties.

4.5. Other Potential Applications and Perspectives of HPH in Wine

High pressure treatments, and HPH in particular, are interesting techniques which might be tested for different winemaking applications, not only for those discussed above.

For instance, the ability of HPH to induce protein unfolding and enzyme inactivation (Sections 3.4 and 3.5) might be exploited for must and wine protein stabilization, as well as for polyphenoloxidase inactivation in grape juice.

Furthermore, when applied on must, HPH has demonstrated to be a suitable technique to increase the dominance of commercial non-*Saccharomyces* yeast strains, when such microorganisms are used in sequential inoculation with *S. cerevisiae* [60,64,65]. This represents an innovative approach in managing wine alcoholic fermentation, reducing the competition by wild microorganisms and allowing the reduction of sulfur dioxide addition. Moreover, this opportunity may be particularly interesting in some specific enological sectors, such as organic winemaking, as well as for the production of SO_2-free wines. Concerning fermentations, HPH and UHPH might also be tested for controlling malolactic fermentation during wine storage, as well as for the inactivation of *Brettanomyces* before barrel ageing.

Finally, it is interesting to mention the work of Serrazanetti et al. [87], who found that HPH affects membrane fatty acid composition by increasing the percentage of unsaturated fatty acids (UFA) when compared with saturated fatty acids (SFA). The role of UFA as a survival factor for yeast metabolism is well known [4] and this observation might open new perspectives in managing yeast nutrition and pre-fermentative operations.

5. Conclusions

Based on the literature currently available, HPH and UHPH seem to have good potential concerning their application to winemaking. Currently, these technologies are not included among the enological practices recommended by the OIV and most of their potential enological applications still need to be tested at lab or pilot-plant scale. However, due to the fast development of HPH technology and its installations in food processing, it can reasonably be assumed that high pressure technologies will raise a great interest in the coming years, also in the wine sector.

Funding: This research received no external funding.

Acknowledgments: The authors are grateful to Enologica Vason S.p.A. for allowing the publication of the data reported in Table 3.

Conflicts of Interest: The authors declare no conflict of interest.

References

1. Corrales, M.; Toepfl, S.; Butz, P.; Knorr, D.; Tauscher, B. Extraction of anthocyanins from grape by-products assisted by ultrasonics, high hydrostatic pressure or pulsed electric fields: A comparison. *Innov. Food Sci. Emerg. Technol.* **2008**, *9*, 85–91. [CrossRef]
2. Morata, A.; Loira, I.; Vejarano, R.; Bañuelos, M.A.; Sanz, P.D.; Otero, L.; Suárez-Lepe, J.A. Grape Processing by High Hydrostatic Pressure: Effect on Microbial Populations, Phenol Extraction and Wine Quality. *Food Bioprocess Technol.* **2015**, *8*, 277–286. [CrossRef]
3. Santos, M.C.; Nunes, C.; Saraiva, J.A.; Coimbra, M.A. Chemical and physical methodologies for the replacement/reduction of sulfur dioxide use during winemaking: Review of their potentialities and limitations. *Eur. Food Res. Technol.* **2012**, *234*, 1–12. [CrossRef]
4. Ribéreau-Gayon, P.; Dubourdieu, D.; Donèche, B.; Lonvaud, A. *Handbook of Enology Vol. 1 The Microbiology of Wine and Vinifications*, 2nd ed.; John Wiley & Sons Ltd.: West Sussex, UK, 2006.
5. Commission Regulation (EC) No. 606 Laying down certain detailed rules for implementing Council Regulation (EC) No 479/2008 as regards the categories of grapevine products, oenological practices and the applicable restrictions. *J. Eur. Commun.* **2009**, *193*, 1–59.
6. International Organization of Vine and Wine, O.I.V. *International Code of Oenological Practices*; O.I.V.: Paris, France, 2018.
7. Walstra, P. Principles of emulsion formation. *Chem. Eng. Sci.* **1993**, *48*, 333–349. [CrossRef]
8. Dumay, E.; Chevalier-Lucia, D.; Picart-Palmade, L.; Benzaria, A.; Gràcia-Julià, A.; Blayo, C. Technological aspects and potential applications of (ultra) high-pressure homogenisation. *Trends Food Sci. Technol.* **2013**, *31*, 13–26. [CrossRef]
9. Floury, J.; Bellettre, J.; Legrand, J.; Desrumaux, A. Analysis of a new type of high pressure homogeniser. A study of the flow pattern. *Chem. Eng. Sci.* **2004**, *59*, 843–853. [CrossRef]
10. Martínez-Monteagudo, S.I.; Yan, B.; Balasubramaniam, V.M. Engineering Process Characterization of High-Pressure Homogenization—from Laboratory to Industrial Scale. *Food Eng. Rev.* **2017**, *9*, 143–169. [CrossRef]
11. Stang, M.; Schauschmann, H.; Schubert, H. Emulsification in high-pressure homogenizers. *Eng. Life Sci.* **2001**, *1*, 151–157. [CrossRef]

12. Calligaris, S.; Plazzotta, S.; Bot, F.; Grasselli, S.; Malchiodi, A.; Anese, M. Nanoemulsion preparation by combining high pressure homogenization and high power ultrasound at low energy densities. *Food Res. Int.* **2016**, *83*, 25–30. [CrossRef]
13. Plazzotta, S.; Manzocco, L. Effect of ultrasounds and high pressure homogenization on the extraction of antioxidant polyphenols from lettuce waste. *Innov. Food Sci. Emerg. Technol.* **2018**, *50*, 11–19. [CrossRef]
14. Balasubramaniam, V.M.B.; Martínez-Monteagudo, S.I.; Gupta, R. Principles and application of high pressure-based technologies in the food industry. *Annu. Rev. Food Sci. Technol.* **2015**, *6*, 435–462. [CrossRef] [PubMed]
15. Zamora, A.; Guamis, B. Opportunities for Ultra-High-Pressure Homogenisation (UHPH) for the Food Industry. *Food Eng. Rev.* **2015**, *7*, 130–142. [CrossRef]
16. McClements, D.J. Edible nanoemulsions: Fabrication, properties, and functional performance. *Soft Matter* **2011**, *7*, 2297–2316. [CrossRef]
17. Silva, H.D.; Cerqueira, M.A.; Vicente, A.A. Nanoemulsions for Food Applications: Development and Characterization. *Food Bioprocess Technol.* **2012**, *5*, 854–867. [CrossRef]
18. Håkansson, A. Emulsion Formation by Homogenization: Current Understanding and Future Perspectives. *Annu. Rev. Food Sci. Technol.* **2019**, *10*, 239–258. [CrossRef] [PubMed]
19. Patrignani, F.; Lanciotti, R. Applications of high and ultra high pressure homogenization for food safety. *Front. Microbiol.* **2016**, *7*, 1132. [CrossRef]
20. Sevenich, R.; Mathys, A. Continuous Versus Discontinuous Ultra-High-Pressure Systems for Food Sterilization with Focus on Ultra-High-Pressure Homogenization and High-Pressure Thermal Sterilization: A Review. *Compr. Rev. Food Sci. Food Saf.* **2018**, *17*, 646–662. [CrossRef]
21. Barba, F.J.; Grimi, N.; Vorobiev, E. New Approaches for the Use of Non-conventional Cell Disruption Technologies to Extract Potential Food Additives and Nutraceuticals from Microalgae. *Food Eng. Rev.* **2014**, *7*, 45–62. [CrossRef]
22. Liu, D.; Ding, L.; Sun, J.; Boussetta, N.; Vorobiev, E. Yeast cell disruption strategies for recovery of intracellular bio-active compounds—A review. *Innov. Food Sci. Emerg. Technol.* **2016**, *36*, 181–192. [CrossRef]
23. Belmiro, R.H.; Tribst, A.A.L.; Cristianini, M. Application of high-pressure homogenization on gums. *J. Sci. Food Agric.* **2018**, *98*, 2060–2069. [CrossRef] [PubMed]
24. dos Santos Aguilar, J.; Cristianini, M.; Sato, H. Modification of enzymes by use of high-pressure homogenization. *Food Res. Int.* **2018**, *109*, 120–125. [CrossRef] [PubMed]
25. Floury, J.; Desrumaux, A.; Lardières, J. Effect of high-pressure homogenization on droplet size distributions and rheological properties of model oil-in-water emulsions. *Innov. Food Sci. Emerg. Technol.* **2000**, *1*, 127–134. [CrossRef]
26. Qian, C.; McClements, D.J. Formation of nanoemulsions stabilized by model food-grade emulsifiers using high-pressure homogenization: Factors affecting particle size. *Food Hydrocoll.* **2011**, *25*, 1000–1008. [CrossRef]
27. Calligaris, S.; Plazzotta, S.; Valoppi, F.; Anese, M. Combined high-power ultrasound and high-pressure homogenization nanoemulsification: The effect of energy density, oil content and emulsifier type and content. *Food Res. Int.* **2018**, *107*, 700–707. [CrossRef]
28. Donsì, F.; Ferrari, G.; Lenza, E.; Maresca, P. Main factors regulating microbial inactivation by high-pressure homogenization: Operating parameters and scale of operation. *Chem. Eng. Sci.* **2009**, *64*, 520–532. [CrossRef]
29. Innocente, N.; Marino, M.; Calligaris, S. Recovery of brines from cheesemaking using High-Pressure Homogenization treatments. *J. Food Eng.* **2019**, *247*, 188–194. [CrossRef]
30. Thiebaud, M.; Dumay, E.; Picart, L.; Guiraud, J.P.; Cheftel, J.C. High-pressure homogenisation of raw bovine milk. Effects on fat globule size distribution and microbial inactivation. *Int. Dairy J.* **2003**, *13*, 427–439. [CrossRef]
31. Wuytack, E.Y.; Diels, A.M.J.; Michiels, C.W. Bacterial inactivation by high-pressure homogenisation and high hydrostatic pressure. *Int. J. Food Microbiol.* **2002**, *77*, 205–212. [CrossRef]
32. Vachon, J.F.; Kheadr, E.E.; Giasson, J.; Paquin, P.; Fliss, I. Inactivation of foodborne pathogens in milk using dynamic high pressure. *J. Food Prot.* **2002**, *65*, 345–352. [CrossRef]
33. Lanciotti, R.; Vannini, L.; Patrignani, F.; Iucci, L.; Vallicelli, M.; Ndagijimana, M.; Guerzoni, M.E. Effect of high pressure homogenisation of milk on cheese yield and microbiology, lipolysis and proteolysis during ripening of Caciotta cheese. *J. Dairy Res.* **2006**, *73*, 216–226. [CrossRef] [PubMed]

34. Tahiri, I.; Makhlouf, J.; Paquin, P.; Fliss, I. Inactivation of food spoilage bacteria and *Escherichia coli* O157:H7 in phosphate buffer and orange juice using dynamic high pressure. *Food Res. Int.* **2006**, *39*, 98–105. [CrossRef]
35. Chaves-López, C.; Lanciotti, R.; Serio, A.; Paparella, A.; Guerzoni, E.; Suzzi, G. Effect of high pressure homogenization applied individually or in combination with other mild physical or chemical stresses on *Bacillus cereus* and *Bacillus subtilis* spore viability. *Food Control* **2009**, *20*, 691–695. [CrossRef]
36. Chen, W.; Harte, F.M.; Davidson, P.M.; Golden, D.A. Inactivation of *Alicyclobacillus acidoterrestris* using high pressure homogenization and dimethyl dicarbonate. *J. Food Prot.* **2013**, *76*, 1041–1045. [CrossRef] [PubMed]
37. Bevilacqua, A.; Corbo, M.R.; Sinigaglia, M. Use of natural antimicrobials and high pressure homogenization to control the growth of *Saccharomyces bayanus* in apple juice. *Food Control* **2012**, *24*, 109–115. [CrossRef]
38. Poliseli-Scopel, F.H.; Hernández-Herrero, M.; Guamis, B.; Ferragut, V. Sterilization and aseptic packaging of soymilk treated by ultra high pressure homogenization. *Innov. Food Sci. Emerg. Technol.* **2014**, *22*, 81–88. [CrossRef]
39. Calligaris, S.; Foschia, M.; Bartolomeoli, I.; Maifreni, M.; Manzocco, L. Study on the applicability of high-pressure homogenization for the production of banana juices. *LWT Food Sci. Technol.* **2012**, *45*, 117–121. [CrossRef]
40. Maresca, P.; Donsì, F.; Ferrari, G. Application of a multi-pass high-pressure homogenization treatment for the pasteurization of fruit juices. *J. Food Eng.* **2011**, *104*, 364–372. [CrossRef]
41. Dong, X.; Zhao, M.; Shi, J.; Yang, B.; Li, J.; Luo, D.; Jiang, G.; Jiang, Y. Effects of combined high-pressure homogenization and enzymatic treatment on extraction yield, hydrolysis and function properties of peanut proteins. *Innov. Food Sci. Emerg. Technol.* **2011**, *12*, 478–483. [CrossRef]
42. Tao, X.; Cai, Y.; Liu, T.; Long, Z.; Huang, L.; Deng, X.; Zhao, Q.; Zhao, M. Effects of pretreatments on the structure and functional properties of okara protein. *Food Hydrocoll.* **2019**, *90*, 394–402. [CrossRef]
43. Cho, S.-C.; Choi, W.-Y.; Oh, S.-H.; Lee, C.-G.; Seo, Y.-C.; Kim, J.-S.; Song, C.-H.; Kim, G.-V.; Lee, S.-Y.; Kang, D.-H.; et al. Enhancement of lipid extraction from marine microalga, *Scenedesmus* associated with high-pressure homogenization process. *J. Biomed. Biotechnol.* **2012**, *2012*. [CrossRef] [PubMed]
44. Augusto, P.E.D.; Ibarz, A.; Cristianini, M. Effect of high pressure homogenization (HPH) on the rheological properties of a fruit juice serum model. *J. Food Eng.* **2012**, *111*, 474–477. [CrossRef]
45. Bot, F.; Calligaris, S.; Cortella, G.; Nocera, F.; Peressini, D.; Anese, M. Effect of high pressure homogenization and high power ultrasound on some physical properties of tomato juices with different concentration levels. *J. Food Eng.* **2017**, *213*, 10–17. [CrossRef]
46. Venir, E.; Marchesini, G.; Biasutti, M.; Innocente, N. Dynamic high pressure-induced gelation in milk protein model systems. *J. Dairy Sci.* **2010**, *93*, 483–494. [CrossRef] [PubMed]
47. Song, X.; Zhou, C.; Fu, F.; Chen, Z.; Wu, Q. Effect of high-pressure homogenization on particle size and film properties of soy protein isolate. *Ind. Crops Prod.* **2013**, *43*, 538–544. [CrossRef]
48. Liu, H.-H.; Kuo, M.-I. Ultra high pressure homogenization effect on the proteins in soy flour. *Food Hydrocoll.* **2016**, *52*, 741–748. [CrossRef]
49. Chen, X.; Zhou, R.; Xu, X.; Zhou, G.; Liu, D. Structural modification by high-pressure homogenization for improved functional properties of freeze-dried myofibrillar proteins powder. *Food Res. Int.* **2017**, *100*, 193–200. [CrossRef]
50. Yu, C.; Cha, Y.; Wu, F.; Xu, X.; Qin, Y.; Li, X.; Du, M. Effects of high-pressure homogenisation on structural and functional properties of mussel (*Mytilus edulis*) protein isolate. *Int. J. Food Sci. Technol.* **2018**, *53*, 1157–1165. [CrossRef]
51. Panozzo, A.; Manzocco, L.; Calligaris, S.; Bartolomeoli, I.; Maifreni, M.; Lippe, G.; Nicoli, M.C. Effect of high pressure homogenisation on microbial inactivation, protein structure and functionality of egg white. *Food Res. Int.* **2014**, *62*, 718–725. [CrossRef]
52. Villay, A.; de Filippis, F.; Picton, L.; Le Cerf, D.; Vial, C.; Michaud, P. Comparison of polysaccharide degradations by dynamic high-pressure homogenization. *Food Hydrocoll.* **2012**, *27*, 278–286. [CrossRef]
53. Le Thanh-Blicharz, J.; Lewandowicz, G.; Błaszczak, W.; Prochaska, K. Starch modified by high-pressure homogenisation of the pastes - Some structural and physico-chemical aspects. *Food Hydrocoll.* **2012**, *27*, 347–354. [CrossRef]
54. Floury, J.; Desrumaux, A.; Axelos, M.A.V.; Legrand, J. Degradation of methylcellulose during ultra-high pressure homogenisation. *Food Hydrocoll.* **2002**, *16*, 47–53. [CrossRef]

55. Alvarez-Sabatel, S.; Marañón, I.M.D.; Arboleya, J.-C. Impact of high pressure homogenisation (HPH) on inulin gelling properties, stability and development during storage. *Food Hydrocoll.* **2015**, *44*, 333–344. [CrossRef]
56. Bot, F.; Calligaris, S.; Cortella, G.; Plazzotta, S.; Nocera, F.; Anese, M. Study on high pressure homogenization and high power ultrasound effectiveness in inhibiting polyphenoloxidase activity in apple juice. *J. Food Eng.* **2018**, *221*, 70–76. [CrossRef]
57. Delfini, C.; Conterno, L.; Carpi, G.; Rovere, P.; Tabusso, A.; Cocito, C.; Amati, A. Microbiological stabilisation of grape musts and wines by high hydrostatic pressures. *J. Wine Res.* **1995**, *6*, 143–151. [CrossRef]
58. Corrales, M.; García, A.F.; Butz, P.; Tauscher, B. Extraction of anthocyanins from grape skins assisted by high hydrostatic pressure. *J. Food Eng.* **2009**, *90*, 415–421. [CrossRef]
59. Morata, A.; Loira, I.; Vejarano, R.; González, C.; Callejo, M.J.; Suárez-Lepe, J.A. Emerging preservation technologies in grapes for winemaking. *Trends Food Sci. Technol.* **2017**, *67*, 36–43. [CrossRef]
60. Bañuelos, M.A.; Loira, I.; Escott, C.; Del Fresno, J.M.; Morata, A.; Sanz, P.D.; Otero, L.; Suárez-Lepe, J.A. Grape Processing by High Hydrostatic Pressure: Effect on Use of Non-*Saccharomyces* in Must Fermentation. *Food Bioprocess Technol.* **2016**, *9*, 1769–1778. [CrossRef]
61. Mok, C.; Song, K.T.; Park, Y.S.; Lim, S.; Ruan, R.; Chen, P. High hydrostatic pressure pasteurization of red wine. *J. Food Sci.* **2006**, *71*, M265–M269. [CrossRef]
62. Santos, L.M.R.; Oliveira, F.A.; Ferreira, E.H.R.; Rosenthal, A. Application and possible benefits of high hydrostatic pressure or high-pressure homogenization on beer processing: A review. *Food Sci. Technol. Int.* **2017**, *23*, 561–581. [CrossRef]
63. Puig, A.; Vilavella, M.; Daoudi, L.; Guamis, B.; Minguez, S. Microbiological and biochemical stabilization of wines using the high pressure technique. *Bull. OIV* **2003**, *869*, 569–617.
64. Loira, I.; Morata, A.; Bañuelos, M.A.; Puig-Pujol, A.; Guamis, B.; González, C.; Suárez-Lepe, J.A. Use of Ultra-High Pressure Homogenization processing in winemaking: Control of microbial populations in grape musts and effects in sensory quality. *Innov. Food Sci. Emerg. Technol.* **2018**, *50*, 50–56. [CrossRef]
65. Puig, A.; Olmos, P.; Quevedo, J.M.; Guamis, B.; Mínguez, S. Microbiological and Sensory Effects of Musts Treated by High-pressure Homogenization. *Food Sci. Technol. Int.* **2008**, *14*, 5–11. [CrossRef]
66. Comuzzo, P.; Calligaris, S.; Iacumin, L.; Ginaldi, F.; Voce, S.; Zironi, R. Application of multi-pass high pressure homogenization under variable temperature regimes to induce autolysis of wine yeasts. *Food Chem.* **2017**, *224*, 105–113. [CrossRef] [PubMed]
67. Comuzzo, P.; Calligaris, S.; Iacumin, L.; Ginaldi, F.; Palacios Paz, A.E.; Zironi, R. Potential of high pressure homogenization to induce autolysis of wine yeasts. *Food Chem.* **2015**, *185*, 340–348. [CrossRef]
68. Campos, F.P.; Cristianini, M. Inactivation of *Saccharomyces cerevisiae* and *Lactobacillus plantarum* in orange juice using ultra high-pressure homogenisation. *Innov. Food Sci. Emerg. Technol.* **2007**, *8*, 226–229. [CrossRef]
69. Franchi, M.A.; Tribst, A.A.L.; Cristianini, M. High-pressure homogenization: A non-thermal process applied for inactivation of spoilage microorganisms in beer. *J. Inst. Brew.* **2013**, *119*, 237–241. [CrossRef]
70. Lee, M.G.; Ham, T.H.; Song, S.H.; Chung, D.H.; Yoon, W.B. Effects of the high pressure homogenization on the viability of yeast cell and volatile components in non-pasteurized rice wine. *Food Sci. Biotechnol.* **2016**, *25*, 1073–1080. [CrossRef]
71. Charpentier, C.; Feuillat, M. Yeast autolysis. In *Wine Microbiology and Biotechnology*; Fleet, G.H., Ed.; Taylor and Francis: NY, New York, USA, 1993; pp. 225–242.
72. Siddiqi, S.F.; Titchener-Hooker, N.J.; Shamlou, P.A. High pressure disruption of yeast cells: The use of scale down operations for the prediction of protein release and cell debris size distribution. *Biotechnol. Bioeng.* **1997**, *55*, 642–649. [CrossRef]
73. Hetherington, P.J.; Follows, M.; Dunhill, P.; Lilly, M.D. Release of protein from baker's yeast (*Saccharomyces cerevisiae*) by disruption in an industrial homogeniser. *Trans. Inst. Chem. Eng.* **1971**, *49*, 142–148.
74. Follows, M.; Hetherington, P.J.; Dunnill, P.; Lilly, M.D. Release of enzymes from bakers' yeast by disruption in an industrial homogenizer. *Biotechnol. Bioeng.* **1971**, *13*, 549–560. [CrossRef] [PubMed]
75. Patrignani, F.; Ndagijimana, M.; Vernocchi, P.; Gianotti, A.; Riponi, C.; Gardini, F.; Lanciotti, R. High-Pressure homogenization to modify yeast performance for sparkling wine production according to traditional methods. *Am. J. Enol. Vitic.* **2013**, *64*, 258–267. [CrossRef]

76. Shynkaryk, M.V.; Lebovka, N.I.; Lanoisellé, J.L.; Nonus, M.; Bedel-Clotour, C.; Vorobiev, E. Electrically-assisted extraction of bio-products using high pressure disruption of yeast cells (*Saccharomyces cerevisiae*). *J. Food Eng.* **2009**, *92*, 189–195. [CrossRef]
77. González-Marco, A.; Ancín-Azpilicueta, C. Influence of lees contact on evolution of amines in Chardonnay wine. *J. Food Sci.* **2006**, *71*, 544–548. [CrossRef]
78. Carrano, G. *Gestione Dell'elevage Sur Lies Mediante Applicazione di Tecnologie Basate Sull'alta Pressione*; University of Udine: Udine, Italy, 2016.
79. Pozo-Bayón, M.Á.; Andújar-Ortiz, I.; Moreno-Arribas, M.V. Scientific evidences beyond the application of inactive dry yeast preparations in winemaking. *Food Res. Int.* **2009**, *42*, 754–761. [CrossRef]
80. International Organization of Vine and Wine, O.I.V. *International Oenological Codex*; O.I.V.: Paris, France, 2018.
81. Comuzzo, P.; Tat, L.; Tonizzo, A.; Battistutta, F. Yeast derivatives (extracts and autolysates) in winemaking: Release of volatile compounds and effects on wine aroma volatility. *Food Chem.* **2006**, *99*, 217–230. [CrossRef]
82. Pozo-Bayón, M.Á.; Andújar-Ortiz, I.; Moreno-Arribas, M.V. Volatile profile and potential of inactive dry yeast-based winemaking additives to modify the volatile composition of wines. *J. Sci. Food Agric.* **2009**, *89*, 1665–1673. [CrossRef]
83. Santos, M.C.; Nunes, C.; Ferreira, A.S.; Jourdes, M.; Teissedre, P.-L.; Rodrigues, A.; Amado, O.; Saraiva, J.A.; Coimbra, M.A. Comparison of high pressure treatment with conventional red wine aging processes: Impact on phenolic composition. *Food Res. Int.* **2019**, *116*, 223–231. [CrossRef]
84. Talcott, S.T.; Brenes, C.H.; Pires, D.M.; Del Pozo-Insfran, D. Phytochemical stability and color retention of copigmented and processed muscadine grape juice. *J. Agric. Food Chem.* **2003**, *51*, 957–963. [CrossRef]
85. Santos, M.C.; Nunes, C.; Rocha, M.A.M.; Rodrigues, A.; Rocha, S.M.; Saraiva, J.A.; Coimbra, M.A. High pressure treatments accelerate changes in volatile composition of sulphur dioxide-free wine during bottle storage. *Food Chem.* **2015**, *188*, 406–414. [CrossRef]
86. Santos, M.C.; Nunes, C.; Rocha, M.A.M.; Rodrigues, A.; Rocha, S.M.; Saraiva, J.A.; Coimbra, M.A. Impact of high pressure treatments on the physicochemical properties of a sulphur dioxide-free white wine during bottle storage: Evidence for Maillard reaction acceleration. *Innov. Food Sci. Emerg. Technol.* **2013**, *20*, 51–58. [CrossRef]
87. Serrazanetti, D.I.; Patrignani, F.; Russo, A.; Vannini, L.; Siroli, L.; Gardini, F.; Lanciotti, R. Cell membrane fatty acid changes and desaturase expression of *Saccharomyces bayanus* exposed to high pressure homogenization in relation to the supplementation of exogenous unsaturated fatty acids. *Front. Microbiol.* **2015**, *6*, 1–10. [CrossRef] [PubMed]

Review

Thermal and Non-Thermal Physical Methods for Improving Polyphenol Extraction in Red Winemaking

Marcos Maza [1,2], Ignacio Álvarez [2] and Javier Raso [2,*]

1 Departamento de Ciencias Enológicas y Agroalimentarias, Facultad de Ciencias Agrarias, Universidad Nacional de Cuyo, M5528AHB Mendoza, Argentina

2 Tecnología de los Alimentos, Facultad de Veterinaria, Instituto Agroalimentario de Aragón-IA2, Universidad de Zaragoza-CITA, c/Miguel Servet, 177, 50013 Zaragoza, Spain

* Correspondence: jraso@unizar.es; Tel.: +34-9-7676-2675

Received: 10 May 2019; Accepted: 18 July 2019; Published: 1 August 2019

Abstract: Maceration-fermentation is a critical stage in the elaboration of high-quality red wine. During this stage, the solid parts of the grape berries remain in contact with the fermenting must in order to extract polyphenols mainly located in the grape skin cells. Extracted polyphenols have a considerable impact on sensory properties (color, flavor, astringency, and bitterness) and on the aging behavior of red wine. In order to obtain wines with a sufficient proportion of those compounds, long maceration times are required. The presence of the solid parts of the grapes during red wine fermentation involves several problems for the wineries such as production capacity reduction, higher energy consumption for controlling the fermentation temperature and labor and energy consumption for periodically pump the grape must over the skin mass. Physical techniques based on heating such as thermovinification and flash expansion are currently being applied in wineries to improve the extraction of polyphenols and to reduce maceration time. However, these techniques present a series of problems derived from the heating of the grapes that affect wine quality. A series of recent studies have demonstrated that non-thermal innovative technologies such as pulsed electric fields (PEF) and ultrasound may represent effective alternatives to heating for assisting polyphenol extraction. In terms of general product quality and energetic requirements, this review compares these thermal and non-thermal physical technologies that aim to reduce maceration time.

Keywords: red wine; thermovinification; flash-release; pulsed electric fields; ultrasound

---

1. Introduction

Red wine is obtained from the must of red grapes that undergoes fermentation together with the solid parts of the grape berries. In this step, known as maceration-fermentation, sugars of the must are converted into ethanol by yeast, and polyphenolic compounds are extracted mainly from the grape skin and the seeds.

Maceration-fermentation is the most critical stage in the red winemaking process. It is essential for obtaining high quality red wines, but is also the one that requires the most energy and workforce. It is estimated that about 64.3% of the total energy needed to produce a liter of wine is consumed during the maceration-fermentation stage [1]. Polyphenols are key actors in red wine, since they are involved in its sensory properties (color, flavor, astringency, and bitterness) [2], in its aging behavior, and in beneficial health effects attributed to moderate wine consumption [3]. In traditional red winemaking, in order to obtain a final product with high polyphenol content, the solid parts of the grape pomace remain in contact with the must during the entire alcoholic fermentation process (7–10 days), or even over a longer period of time. Although maximum anthocyanin content and color intensity is already achieved during the first days of maceration [4,5], the extraction of procyanidins and other flavonoids, which have significant impact on other sensory attributes such as astringency and mouthfeel, requires longer maceration periods [6,7]. As these compounds are mainly located in the seeds, its extraction required the presence of ethanol to disorganize the outer lipidic cuticle surrounding the seeds [8]. On the other hand, in red winemaking: aromatic precursors responsible for the varietal aromas in wines are extracted from the solid parts of grape barriers, along with polyphenolic compounds.

The necessity of maintaining the solid parts of the grape berries in contact with the fermenting must leads to several issues faced by wineries in the red winemaking process [9]. It is estimated that approximately 20% of the fermentation tanks are occupied by the solid parts, resulting in a reduction of the effective volume of the tanks and, as a consequence, of a winery's production capacity. This issue becomes especially significant at the peak of harvesting, when the fermentation-maceration tanks' production capacity may be exceeded. Other negative side effects of longer maceration periods are related with the difficulty of controlling the temperature increment as a consequence of the fermenting activity of the yeasts when the solid parts are present in the fermentation tanks, as well as with the labor force and energy consumption required to periodically pump the wine over the skin mass that rises to the top of the fermentation tanks [10].

Different strategies have been adopted in wineries to enhance the extraction of phenolic compounds and to reduce the duration of the maceration-fermentation stage in red winemaking [11,12]. Physical technologies based on heating, such as thermovinification and flash expansion, are currently being applied in wineries for this purpose [13]. They present a series of problem such as the difficulty involved in stabilizing the color, the loss of varietal aromas through temperature increment, and the consumption of high quantities of energy [14,15]. A series of studies have recently demonstrated that non-thermal innovative technologies such as pulsed electric fields and ultrasound may represent effective alternatives to heating in the attempt to improve polyphenol extraction [16–19]. This review compares thermal and non-thermal physical technologies that aim to reduce maceration time in terms of equipment complexity, energetic requirements, and overall quality of the red wine.

2. Thermal Technologies for Improving Polyphenol Extraction

Although the heating of red grapes in order to reduce maceration has been investigated since the early 20th century, the process was not commercially adopted until the 1970s, when industrial heating systems were developed for that purpose [20].

In general terms, the process consists in heating grapes to over 70 °C for a period of time ranging from a few minutes to several hours. As a consequence of heating, the cell envelopes of the grape skins are braked down, thereby facilitating the subsequent release of polyphenols (mainly anthocyanins) that are located inside the cells into the liquid phase [21]. Heating also denatures enzymes such as polyphenol oxidase, thereby preventing browning. In fact, heating was originally used to prevent laccase activity in grapes contaminated with the mold *Botrytis cinerea* [22].

Although generally heating of the grapes before fermentation is called "thermovinification", different pre-fermentation heating processes are currently being applied in wineries. These techniques can be classified into two groups, depending on whether the cooling of the grapes, similarly to heating, is conducted using heat exchangers, or whether the cooling is conducted into a vacuum chamber. The first kind of process is designated as "thermovinification", along with its variations, known as "pre-fermentation hot maceration" (MPC), and "short-time-high-temperature treatment with warm maceration" (KZHE). The second group involves the technique called thermo-flash, flash détente or flash-release.

2.1. *Thermovinification, MPC, and KZHE*

2.1.1. Description of the Techniques

Thermovinification, MPC, and KZHE are pre-fermentative heating techniques; they all have in common that the temperature of the grape mash does not increase above 85 °C, and that heating and cooling are conducted in heat exchangers [23].

In thermovinification, heating up to around 70 °C is conducted for a period of time of less than one hour, after which the grape mash is pressed to separate the solid parts and perform fermentation as for white wine. If heating at the same temperature is extended for a longer period of time (up to 24 h), and the fermentation is conducted in the presence or absence of the solid phase, the process is called MPC ("pre-fermentation hot maceration"). A variation of MPC is the process developed in Germany called KZHE ("short-time-high-temperature treatment with warm maceration"). In the latter, fermentation is conducted in the absence of solids after maintaining the grapes at around 45 °C for 6–10 h after having heated them to around 85 °C for 2 min.

2.1.2. Equipment

The simplest and most inexpensive heat exchangers used to heat grapes before fermentation are tube-in-tube heat exchangers. To prevent blocking problems in this heat exchangers, it is required the application of the treatment to the entire mix of juice and solid parts. To save energy, it is recommended to treat the solid parts after pre-draining in order to minimize the quantity of material that needs to be heated and cooled. In this case, it is recommended to use a scraped-surface heat exchanger with a rotating shaft that improves heat transfer to the product. This approach permits to process the grape mash with a moderate degree of pre-draining while avoiding blocking issues.

Different approaches have been developed to save energy in the heating of the grape mash by recovering heat. In such systems, incoming well-mixed crushed grapes without any pre-draining are pre-heated together with the crushed grapes that have already been heated. In these systems, and in order to avoid blocking, spiral heat exchangers or heat exchangers with a section of rectangular or parallel rectangular channels are preferred.

An alternative to the above-described continuous single pass method is to heat the grape mash with a tube-in-tube heat exchanger while recirculating them on a tank. This approach, generally used in smaller wineries, results in slower and more heterogeneous heating.

For transformation the sugar of must into ethanol by yeasts during fermentation, temperatures between 20 and 30 °C are required. Therefore, after the heating period, it is necessary to cool down the grape mass prior to fermentation. The cooling step is conducted with heat exchangers similar to those that are used for heating.

Fluids used in this type of equipment are hot water or steam for heating, and cold water or glycol for cooling.

In general, such installations used for pre-fermentative heating occupy a considerable area within the winery. The space is required for the heat exchanger systems as well as for the facilities designed to heat and cool the fluids.

2.1.3. Impact of the Treatment in the Composition of Wine

The main objective in using these pre-fermentation heating techniques is to speed up the extraction of polyphenols from the grape skins with the purpose of eliminating or reducing the maceration stage. However, the characteristics of the final wine obtained with such heated grapes may be affected [24,25].

As a consequence of heating, wild yeast populations are inactivated, thus requiring the addition of microbial starters to trigger fermentation. Generally, alcoholic fermentation is initiated without problems after pre-fermentation heating. Occasionally a more abrupt fermentation than in traditional fermentation is observed, probably related to the release of nutrients from the solid parts of the grapes as a consequence of heating [26]. A significant increase in sugar concentration, pH, amino acids, and ammonium in thermovinified Carignan must was reported [27]. Bacterial populations of lactic as well as acetic bacteria are also inactivated, resulting in wines with low volatile acid content. Total acidity of wine is not usually affected by pre-fermentation heating. Although a more elevated extraction of cations and anions as a consequence of grape heating has been described, they precipitate as salts of tartaric acid, thus ultimately leaving wines thus obtained in the same condition as untreated wines [28].

Pre-fermentation heating, in which the solid parts of the grapes are pressed and fermentation is conducted in the liquid phase, has the main objective of enhancing the extraction of color from the skins. The color increment is a consequence of the rapid extraction of anthocyanins.

While anthocyanins are extracted since the first moments of fermentation, flavanols require the presence of ethanol to be extracted.

Piccardo and González-Neves [29] reported that the extraction of anthocyanins after thermovinification was practically immediate. As consequence the anthocyanin concentration and the color intensity in the first days of fermentation were 21% and 45% higher, respectively, than in control. Most studies of the thermovinification technique have been conducted with Pinot noir due to the difficulty of extracting anthocyanins from that grape variety. It has been reported that the anthocyanin quantity in the Pinot noir variety reached a maximum at the onset of fermentation, with a concentration 2 to 3 times higher than in traditional fermentation. A drastic decrease in anthocyanins was observed, however, towards the end of fermentation [30]. Studies conducted at laboratory scale have demonstrated the degradation of anthocyanins due to temperature [27,31]. Anthocyanin content was affected by thermovinification when the treatment was very prolonged, or above 70 °C.

Concerning the effect of pre-fermentation heating on aroma, it has been reported that wines have a standardized sensory profile often described by oenologists as "banana yogurt" [32]. For example, varietal aromatic compounds with green pepper aromas (methoxypyrazines) decreased in Cabernet Sauvignon wines when they were thermo-treated [33]. Geffroy et al. [31] reported that a heat treatment at 70 °C for two hours induced a significant loss of several grape-derived aroma compounds (terpenols, norisoprenoids and some phenols) associated with an increase in α-terpineol, guaiacol and 2,6-dimethoxyphenol, suggesting thermal degradation. When thermovinification was applied to Carignan wine at two different temperature levels, 50 °C and 75 °C, and within two different time intervals, 30 min and 3 h, the effect of temperature on aroma composition was greater than that of heating time. Wines obtained from grapes treated at 50 °C had higher concentrations of geraniol, β-citronellol, β-damascenone, and 3-mercaptohexanol, in most cases [27].

Although thermovinification reinforces anthocyanin extraction, the wines thereby obtained are known to lack color stability and structure. Anthocyanins can decrease due to enzymatic hydrolysis [34], to combination with proteins, or to re-fixation with solid parts such as the skin [35] and yeasts [36]. Since no alcohol is present at the time of heating, the wine does not contain sufficient levels of tannin to stabilize unstable anthocyanins and to provide structure. As a consequence, wines obtained by thermovinification are not usually used for aging, but commercialized as table wine for everyday use.

Finally, since tannin extraction is much more dependent on increasing ethanol content to encourage its solubilization, one approach to obtain a higher extraction of polyphenolic compounds consists in fermenting grapes after heating with solid parts of the grapes, as in standard vinification with shorter maceration time. This alternative was found to increase total phenolic index, color intensity and anthocyanins content in wine 58%, 25% and 45%, respectively [29].

2.2. Flash Release

2.2.1. Description of the Technique

The process called "flash release" or "flash détente" consists in rapidly heating the grapes at temperatures between 85–95 °C by a direct injection of steam. Grapes are then introduced into a vacuum that instantly vaporizes the water, thereby cooling the treated grapes and weakening their skin cell envelopes by boiling the water inside the cells [37]. This effect on the skin cells enhances extractability in subsequent fermentation process that may be conducted with or without the solid parts of the grapes. A modification of this process is called "half" flash détente [38]. It uses a weaker vacuum to cool the grape mash to around 50 °C instead of 30 °C.

2.2.2. Equipment

Flash release or flash expansion equipment consists of a heat exchanger and a vacuum chamber. In the heat exchanger, the steam is directly injected to the grape mash. Grape mash is continuously moved by two hollow stem augers through which the steam enters into the vacuum chamber. Since the chamber is under negative pressure (20–25 hPa), the water instantly evaporates, while the grape mash is simultaneously cooled. The estimated amount of evaporated water ranges between 6 to 10% [39]. It is condensed in a condenser connected with the vacuum chamber, and reincorporated into the grape mash totally or partially, depending on the amount of water in a gaseous state added to the grape mash during the heating process. The flash release system requires a boiler to produce water vapor for rapid heating.

2.2.3. Impact of the Treatment in the Composition of Wine

It has been reported that the yeast population lag phase before starting fermentation is slightly shorter when the grape mash is treated by flash release, probably because the treatment has triggered the release of some yeast nutrients [40].

Characteristics of wines obtained by flash release can be modulated by conducting fermentation in liquid phase, or by keeping the solid parts of the grapes in contact with the liquid phase for different periods of time. It has been observed that flash release increases the extraction of flavanols and flavonols from skins rather than from seeds. Therefore, when fermentation is carried out without the skins, the concentration of tannins with respect to anthocyanins is low, as in wines obtained via traditional pre-fermentation heating. The destabilization of grape skin cell envelopes seems to facilitate the extraction of tannins located in the vacuoles of the hypodermal cells of the grape skins. However, the proportion of those tannins in the resulting wine is low compared with the tannins coming from the seeds, which require the presence of ethanol to be extracted and also a more maceration time [41].

Morel-Salmi et al. [13] investigated the phenolic extraction kinetics during the maceration-fermentation of Grenache must previously treated by flash release. They observed that the amount of various families of phenolic compounds was higher at the beginning of the fermentation process in the flash release treated must than in control. On the other hand, while the levels of catechins, flavonols, and proanthocyanidins increased during fermentation of flash release treated musts, the concentration of hydroxycinnamic acids remained constant and anthocyanins decreased during the first day, and then they remained constant. The increment in concentration of galloylated units increased throughout fermentation, reflecting the gradual extraction of seed tannins as the ethanol level increased. Therefore, although the effect of flash-release on grape skin cell envelopes is more drastic than that of other pre-fermentation heating techniques, a contact period of the solid parts of the grapes with the must during fermentation after treatment is required in order to obtain structured wines with large amounts of polyphenols. At the of the vinification process, the wine obtained with Grenache grapes treated by flash release had a total phenolic index and a colour intensity 14% and 9% higher than the control wine respectively.

The effect of flash release on the extraction of aromatic compounds and aroma precursors has been also investigated [42]. As compared to wines obtained by other pre-fermentation heating techniques, wines obtained with flash release maintain their varietal aromatic profile. The treatment increases the levels of fatty acid ethyl esters and β-ionone in Grenache wines. On the other hand, it has been observed that flash release may reduce the content of C6 compounds responsible for herbaceous aromas [43]. This effect is especially interesting when the wines are elaborated with grapes that have not reached their optimal stage of maturity.

Wines of different varieties such as Grenache, Carignan, Syrah, and Mourvedre obtained with flash expansion technique were preferred to control wines in a sensory analysis, especially when the contact time of the solid parts of the grapes with the fermenting must was extended [44].

3. Non-Thermal Techniques for Improving Polyphenol Extraction

Non-thermal technologies have been one of the most frequently investigated topics in the field of food processing over the last decades [45]. The "non-thermal" concept refers to a group of technologies whose effects in foods are similar to those caused by heating, albeit at temperatures lower than the ones used in thermal processing. Some of these treatments may involve heat due to the generation of internal energy (e.g., resistive heating during PEF). However, they are classified as non-thermal, because they can eliminate or significantly reduce the application of high temperatures in food processing, thereby avoiding the deleterious effects of heat on the flavor, color, and nutritive value of foods.

The emergence of non-thermal technologies can lead to high quality products while saving energy by improving heating efficiency. Most of these technologies are locally clean processes and therefore appear to be more environment-friendly, with less environmental impact than traditional ones [46]. Novel processing technologies are increasingly attracting the attention of food processors, since they can provide food products with improved quality and a reduced environmental footprint, while reducing processing costs and improving the products' added value.

Due to their special mechanism of action, pulsed electric fields and high-intensity ultrasound are among the non-thermal technologies that have been most investigated with the purpose of improving polyphenol extraction in wineries.

3.1. Pulsed Electric Fields (PEF)

3.1.1. Description of the Technique

PEF processing consists in the intermittent application of short duration pulses (ms-µs) of high voltage (kV) to a product located between two electrodes. The applied external voltage generates an electric field whose strength depends not only on voltage intensity, but also on the distance between the electrodes. When exposed to a sufficiently strong electric field, the cell membrane undergoes a phenomenon called electroporation, consisting in the increment of cell envelope permeability as a consequence of the formation of pores in the cytoplasmatic membrane [47].

If the intensity of the electric field is not high enough, or if the exposure to the electric field is sufficiently brief, the membrane can spontaneously return to its initial state and remains viable (reversible electroporation). However, intense electric fields or longer exposures can cause irreversible electroporation [48]. Reversible electroporation is a procedure that is typically used in molecular biology and in clinical biotechnological applications to gain access to the cytoplasm for the introduction or delivery in vivo of drugs, oligonucleotides, antibodies, plasmids, etc. However, the main applications of PEF in the food industry aim to cause irreversible electroporation of the cell membranes. It has been demonstrated that irreversible modification of the permeability of cell membranes can inactivate vegetative cells of microorganisms, enhance mass transfer in different operations of the food industry (e.g., extraction of intracellular components of interest, dehydration, infusion of compounds into the cells, etc.), and modify food structure [49,50].

3.1.2. Equipment

Basic components of an apparatus for the application of PEF are a pulse generator and a treatment chamber. The pulse generator is a Marx generator of square waveform pulses with a direct current power supply which converts alternating current to direct current line that is used, in turn, to charge a set of capacitors at high voltage. When the high voltage switch (a high-power solid-state switch) is opened, the capacitors are charged. If the high-power switch is then closed, all the electrical energy stored in the capacitors is delivered to the treatment chamber. The switching system permits the controlled discharge of the capacitor in the form of pulses of very short duration at very high frequencies (reaching hundreds of pulses per second).

During PEF processing, a liquid food or pumpable product is passed through a treatment chamber where it is subjected to short pulses of high voltage. The treatment chamber consists of two electrodes made of a conducting material such as stainless steel or titanium; they are separated by an insulating material, which forms an enclosure containing the food material. Different types of treatment chambers have been designed to minimize the effect of electrolysis as well as corrosion. The two most important treatment chamber designs that are presently considered for the commercial application of PEF are parallel electrode and co-linear configurations. The latter configuration is the one habitually used for processing crushed grapes after destemming, with the purpose of electroporating the cytoplasmic membrane of grape skin cells to facilitate the extraction of polyphenols during the maceration-fermentation stage. The co-linear treatment chamber consists of an electrically insulating tube through which the grape mash flows. The electrodes are located in the middle (high voltage) and on either side of the chamber (ground). They consist of two metal pipes that also serve as the entrance and exit for the fluid. The circular section of this co-linear configuration facilitates its installation in winery circulation pipes used to transport crushed and destemmed grapes to the fermentation-maceration tanks (Figure 1) [51].

The lack of reliable and viable industrial-scale equipment has limited the commercial exploitation of PEF in the food industry for many years. However, recent developments in pulse power generators have enabled the design of PEF equipment with characteristics that can meet industrial standards in terms of reliability and workloads [52].

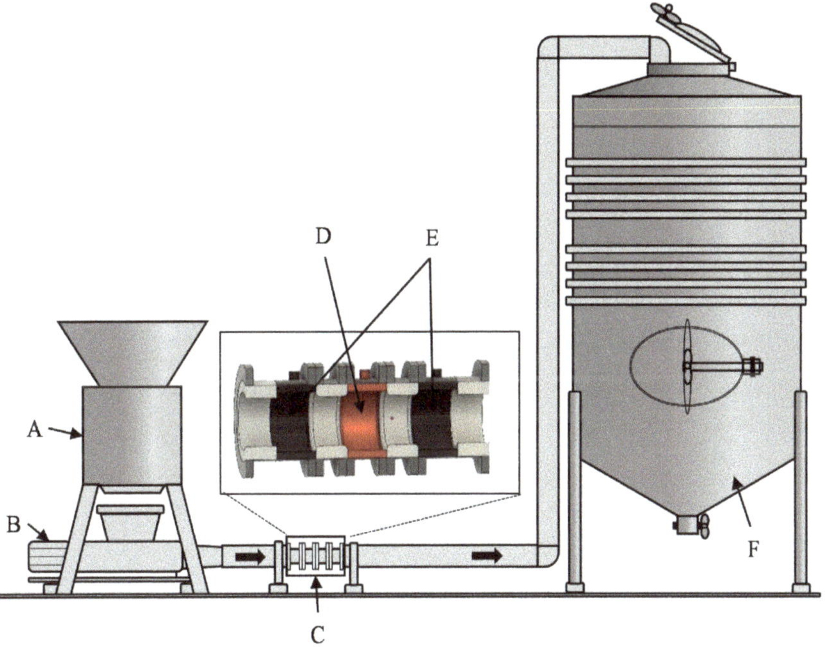

Figure 1. Flow chart of grape processing with PEF technology. (**A**) destemming; (**B**) progressive cavity pump; (**C**) co-linear treatment chamber; (**D**) high voltage electrode; (**E**) ground electrode; (**F**) fermentation tank.

3.1.3. Impact of the Treatment in the Composition of Wine

As compared with heating techniques, of the improvement of extraction of polyphenols by PEF requires to maintain the solid parts of the grapes in contact with the liquid phase for different periods of time [53]. Therefore, the effect of PEF treatment on cell skin envelopes seems to be less aggresive than that of techniques based on heating [54]. Tests carried out by different authors on different grape varieties agree that PEF treatment neither affects the fermentation process nor the physicochemical properties of the resulting red wine. Ethanol content, pH, volatile acidity and total acidity in the wines obtained with grapes treated by PEF were similar to control wines [53,55].

The electroporation of cell grape skins by the application of PEF accelerates and increases the extraction of phenolic compounds during the maceration-fermentation stage in the vinification of red grapes [56]. Different studies have shown that, after the same maceration time than in control wine, PEF treatment reinforces oenological parameters by a rate of 10% to 60%, depending on the extraction of polyphenols (color intensity, total anthocyanin content, and total polyphenol content) in the maceration-fermentation stage [53].

Puértolas et al. [57] showed that PEF technology can help reduce maceration times. Cabernet Sauvignon wine obtained from PEF-treated grapes (5 kV/cm, 150 μs, and 3.67 kJ/kg) presented higher color intensity, total anthocyanin content, and total polyphenol content values, although the duration of the maceration of the grapes treated by PEF was 48 h shorter than for control wines. Evolution during aging of the wine obtained from grapes treated by PEF was similar to control wine. The differences in color intensity, total anthocyanin content, and total polyphenol content observed at the end of fermentation between control wine and the wine obtained from PEF-treated grapes were maintained after aging the wine in bottle or oak barrels [58]. Determination of individual polyphenols by means of high-performance liquid chromatography (HPLC) highlighted that the wines obtained by PEF treatments did not show differences in terms of the proportion of different polyphenols, thus indicating that PEF treatment did not selectively extract phenolic compounds from grape skins. López-Alfaro et al. [59] reported that the content of resveratrol, one of the most researched phenols in wine due to its beneficial properties, increased by a proportion of 200, 60 and 50% in Tempranillo, Garnacha and Graciano, respectively, when the grapes were treated with PEF before maceration-fermentation.

Energetic requirements for the electroporation of cells of grape skins are lower than 10 kJ/kg; as a consequence, the treatment causes an increment of less than 2 °C in grape mash temperature. This low impact allows the obtained wines to maintain their varietal character [57]. Some experiments have shown that PEF treatments encourage the diffusion of aromatic compounds found in the skin, as well as of aromatic precursors [60]. PEF treatment did not increase the concentration of C6 family compounds associated with herbaceous aromas in wines obtained from Garnacha, Tempranillo, and Graciano varieties [60]. The treatment significantly increased monoterpenoid compounds, and a had positive effect on the concentration of β-ionone, total esters, and benzenoid compounds in Grenache wine. However, the volatile composition of Tempranillo and Graciano wines was not affected by PEF.

Sensory analysis did not detect any drawbacks in Cabernet Sauvignon wines obtained with grapes treated by PEF. Luengo et al. [51] compared Grenache wines featuring similar enological parameters in terms of polyphenol content obtained, on the one hand, with PEF treated grapes and 7 days of maceration and, on the other hand, with untreated grapes and 14 days of maceration. Compared with control wine, panelists preferred the wine obtained with grapes treated by PEF and a shorter maceration period.

3.2. *Ultrasound*

3.2.1. Description of the Technique

Acoustic waves of a specific frequency lying above the detection threshold of human hearing (i.e., over 16–18 kHz) are designated as ultrasound. Ultrasound is divided into two categories, according to the frequency range and the intensity of ultrasonic waves. The first group, commonly known as high-intensity ultrasound, features low frequency and high intensity (20–100 kHz; >10 W/cm^2). The second group, commonly called diagnostic ultrasound, uses high frequency and low power (>100 kHz; <1 W/cm^2).

When high-intensity ultrasound passes through a liquid medium, a phenomenon called acoustic cavitation occurs [61]. Cavitation consists in the implosion of bubbles formed in liquid media when the local pressure in the expansion phase falls below vapor pressure. During the implosion, it is estimated that high temperatures and pressures are reached in very small spots and very short periods of time: liquid jets of up to 280 m/s are likewise generated. These phenomena brought about by cavitation are responsible for effects attributed to high-intensity ultrasound, such as the increment of mass transfer, or the breakage of cells of microorganisms, or of plant or animal tissues [62]. Ultrasound may therefore enhance the extraction of polyphenols from the solid parts of grapes in red winemaking by breaking up the cells, and by facilitating the diffusion of polyphenols from the cells to the must [63].

3.2.2. Equipment

An apparatus for the generation of ultrasound consists in a power supply and a transducer. The power supply converts alternating current line voltage to frequencies of over 20 kHz electrical energy. This high-frequency electrical energy is fed to a transducer, where it is converted to mechanical vibrations at the same frequency as the transformed electrical current. The physical concept underlying the transducer is the piezoelectric effect: the property of certain materials causes them to change shape when an electric current is applied to them. An ultrasound transducer contains a thin disk, square, or rectangle of piezoelectric ceramic placed between two electrodes which expand and contract when subjected to alternating voltage. The converter vibrates in a longitudinal direction and transmits the motion to the solution, thereby causing cavitation [61].

A power ultrasound system has recently been developed for processing destemmed and crushed grapes in continuous flow. The equipment consists of a hexagonal stainless-steel pipe into which the transducer is welded (Figure 2). The length of the pipes containing the transducers is variable, depending on the installation's processing capacity, which can reach up to ten tons per hour. The cavitation caused by the ultrasound treatment provokes the destruction of the cells of the solid parts of the grapes, thereby leading to the release of polyphenols.

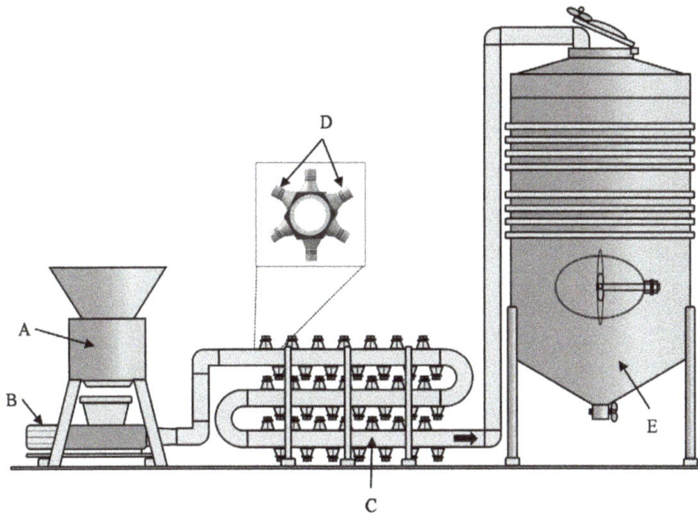

Figure 2. Flow chart of grape processing with ultrasound technology. (**A**) destemming; (**B**) progressive cavity pump; (**C**) ultrasound treatment zone; (**D**) transducer; (**E**) fermentation tank.

3.2.3. Impact of the Treatment in the Composition of Wine

The use of high-power ultrasound (US) to improve the extraction of phenolic compounds from grapes has been recently studied [64,65]. As in the case of PEF technology, an ultrasonic treatment applied at different frequencies (45, 80, and 100 kHz) with the purpose of improving polyphenolic extraction did not modify the physicochemical properties of wine. Total acidity and pH of Cabernet

Sauvignon wine obtained from ultrasound-treated grapes did not show significant differences with respect to control, although electrical conductivity was slightly higher (4%). This increment in conductivity could be associated with the release of ions located inside the cells of the solid parts of the grapes to the must [65].

El Darra et al. [17] investigated the effect of ultrasound on the extraction of polyphenols from Cabernet Sauvignon grapes at laboratory scale using an US probe in a flask containing 400 ± 5 g of must and grape skins. Results showed an increment in the phenolic, anthocyanin, and tannin contents of the wines obtained from grapes treated by ultrasound. A greater color intensity compared with the untreated samples was likewise observed in the wines after ultrasonication treatment, whereby the highest values of those parameters were achieved by the samples that had been subjected to the most intense treatment (363 kJ/kg).

Monastrell wines obtained after different maceration times with grapes treated by a continuous flow pilot-scale power ultrasound system (2500 W, 28 kHz, 8 W/cm^2) were compared with wines obtained from untreated grapes [66]. Results showed an increase in the chromatic characteristics of the wines obtained with ultrasonicated grapes. The values for these chromatic characteristics were higher in wines obtained with ultrasonicated grapes and 3 days of maceration than in control wines with a longer maceration period (5 days). After two months of aging, the wines obtained with grapes treated by US contained between 20 and 35% more total polyphenols than control wines [66]. The ultrasound treatment also encouraged the extraction of tannins from the seeds, although to a lesser extent than tannins from the skins. As a consequence, the wines elaborated with ultrasonicated grapes and 3 days of maceration presented twice the concentration of proanthocyanidins than that of control wines obtained with 8 days of maceration.

Concerning the effect of ultrasonication treatment on the volatile composition of wines, no significant differences were observed in the total concentration of those compounds between control and wine obtained from grapes treated by ultrasound, regardless of maceration time [63].

4. Discussion

Novel non-thermal processing technologies have been developed in the last years with the aim of preventing problems associated with thermal processing, and with the purpose of improving energy efficiency and food production sustainability. The introduction of a new technology on the market requires that it must perform at least as well as existing commercial processes. Table 1 compares, as an example, the improvements derived of application of different thermal and non-thermal physical methods to the grapes before vinification in terms of polyphenolic extraction. It is observed that PEF and ultrasound permits attaining similar enhancements in total anthocyanin content, color intensity and total polyphenol content than techniques based in the heating of the grapes. However, as it is shown in Table 2 thermovinification and flash release present certain drawback related with the wine quality, energy consumption etc. that would support the implementation of non-thermal physical techniques to improve polyphenol extraction.

Although in the past decades the food industry has carried out immense efforts to optimize energy consumption and heat recovery in conventional processes, the introduction of non-thermal technologies may yet provide a further potential to help reduce energy consumption and operational costs while improving food production sustainability. Table 3 compares the energy delivered to grapes (after destemming and crushing) by several thermal (with final treatment temperatures between 50 to 85 °C before fermentation) and non-thermal processes (with temperature increases lying under 5 °C) to obtain an equivalent effect in terms of polyphenol extraction in red winemaking. One can observe that the energy required to increase the temperature of grapes is much higher than the energy required to electroporate grape skin cells by PEF, or to disrupt skin and seed cells by ultrasound. From an energetic point of view, non-thermal techniques present an additional advantage, since the low energy delivered to the product does not substantially increase its temperature. As compared with themovinification or flash release in the case of winemaking, this implies that it is not necessary to waste energy to cool the grape mash to the temperature required to initiate fermentation. According to Table 3, the average specific energy of thermal treatments is 17.6-fold higher than that required for non-thermal processes being the specific energy required for PEF treatment lies 3.2-fold lower than that required by ultrasound treatment. Consequently, considering that the energy source is different for thermal and non-thermal processes, lower operational costs are required for PEF and ultrasound processing. From an energetic point of view, another important issue when comparing thermal and non-thermal technologies is that, in the latter processes, energy is delivered directly to the product, thus making such methods much more efficient than heating techniques where thermal energy is transferred through an intermediate medium (water, water vapor, or oil). While thermal techniques require water, non-thermal techniques permit to obtain similar objectives without increasing water consumption in a winery. As a consequence, non-thermal technologies are considerably more sustainable: they reduce the use of resources as well as CO_2 emissions.

Another aspect that differentiates thermal from non-thermal techniques is related with the installation of the unit in the winery. The required space for the installation of thermovinification or flash expansion is much greater than that required for the installation of ultrasound or PEF units. Generally, considerable renovation is required for a winery to introduce a thermovinification or a flash expansion unit with associated auxiliary units. PEF technology differs from other techniques in view of its portability. The pulse generator unit is separate from the treatment chamber, thereby allowing a rapid adaptation of the process, depending on the product to be treated. Moreover, these units are small enough to be easily integrated into existing production lines without requiring major factory overhaul.

To summarize, non-thermal techniques such as PEF and ultrasound are now increasingly attracting the attention of wineries as an alternative to techniques based on grape heating in order to reduce the duration of maceration time and/or to avoid the purchase of maceration-fermentation tanks. These techniques can encourage the production of wine with improved quality and a reduced environmental footprint, while at the same time decreasing processing costs.

Table 1. Improvements derived of grape treatment before vinification with different thermal and non-thermal physical methods in terms of increment in total polyphenolic content, color intensity and total anthocyanin content.

Technology	Treatment	Variety	Total Polyphenolic Content	Colour Intensity	Total Anthocyanin Content	Ref.
Thermovinification	82 °C 1 h Flow rate: 500 kg/h Maceration time: 5 days	Merlot	36%	N/A	26%	[28]
Flash-release	95 °C for 6 min Strong vacuum (>100 mbar) Flow rate: N/A Maceration time: 5 days	Carignan	11%	30%	30%	[13]
PEF	5 kV/cm, 150 µs (50 pulses 3 µs, 3.67 kJ/kg) Flow rate: 118 kg/h Maceration time: 4 days	Cabernet Sauvignon	23%	38%	34%	[57]
Ultrasound	2500 W; 28 kHz; 8 W/cm^2 Flow rate: 400 kg/h Maceration time: 4 days	Monastrell	32%	31%	13%	[66]

N/A: information not available.

Table 2. Advantages and disadvantages of different thermal and non-thermal technologies for improving polyphenol extraction in red winemaking for grape pre-fermentation treatments.

Technology	Advantage	Disadvantages	Ref.
Thermovinification	Possibility of obtaining red wines without maceration For obtaining table wines. Permits to inactive enzymes and microorganisms Approved by OIV	Poor color stability Possible degradation of anthocyanins Loss of varietal aromatic compounds Wines not usually used for aging Addition of starter cultures for initiating fermentation required High energetic requirement. Supplies of methane or diesel oil required	[12,25,27–29]
Flash-release	Mainly for obtaining table wines. Permits to inactive enzymes and microorganisms Obtaining of more complex sensory characteristics Approved by OIV	Possible degradation of anthocyanins Wines not usually used for aging Addition of starter cultures for initiating fermentation required High energetic requirement. Supplies of methane or diesel oil required Renovations are required in the winery for installation (Large facilities: >100 m^2)	[13,39,44]
PEF	Demonstrated the ability of aging of the wines in oak barrels Easy implementation in the winery (small facilities:<10 m^2) Possibility of renting the PEF unit Possibility of conducting fermentations with wild yeast Possibility of using for other applications in winery (microbial inactivation or accelerating aging on the lees) Low energy requirements	Maceration of few days is required for obtaining red wines Approval for the OIV in process. No enzymatic inactivation.	[16,51,53,58]
Ultrasound	Easy implementation in the winery (small facilities:<10 m^2) Possibility of renting the ultrasound unit Possibility of conducting fermentations with wild yeast Possibility of using for other applications in winery (accelerating aging on the lees) Low energy requirements	Maceration of few days is required for obtaining red wines Approval for the OIV in process. No enzymatic inactivation.	[66–68]

Table 3. Estimation of the energetic costs of thermal and non-thermal physical methods for improving polyphenol extraction during winemaking.

Technology	Specific Energy Delivered to the Grape (kJ/kg)	Additional Specific Energy * (kJ/kg)	Total Specific Energy (kJ/kg)	kWh/tn	€/tn [a]
Thermovinification	161.92	40.68	202.6	56.28	7.32
Flash-release	251.88	45.86	297.7	82.70	10.75
PEF	6.70	-	6.70	1.86	0.24
Ultrasound	21.60	-	21.60	6.0	0.78

[a] Energy cost: Electricity: 0.13€/ kWh, * The energy required for the complete operation of the thermal system (pumps, refrigeration and condensation systems).

Author Contributions: All authors contributed to the writing of the manuscript, read, and approved the final version.

Funding: M.M. is supported by a predoctoral scholarship from the Universidad Nacional de Cuyo, Argentina Res: RE 4974/2016.

Acknowledgments: Thanks go to the European Regional Development Fund, to the Department of Innovation Research and University Education of the Aragon Government, and the European Social Fund (ESF).

Conflicts of Interest: Authors declare no conflict of interest.

References

1. Genc, M.; Genc, S.; Goksungur, Y. Exergy analysis of wine production: Red wine production process as a case study. *Appl. Therm. Eng.* **2017**, *117*, 511–521. [CrossRef]
2. Arnold, R.A.; Noble, A.C. Bitterness and Astringency of Grape Seed Phenolics in a Model Wine Solution. *Am. J. Enol. Vitic.* **1978**, *29*, 150–152.
3. Perez-Vizcaino, F.; Fraga, C.G. Research trends in flavonoids and health. *Arch. Biochem. Biophys.* **2018**, *646*, 107–112. [CrossRef] [PubMed]
4. Boulton, R.; Singleton, V.L.; Bisson, L.F.; Kunkee, R.E. *Principles and Practices of Winemaking*; Springer Science & Business Media: Berlin, Germany, 2013; ISBN 978-1-4757-6255-6.
5. Zanoni, B.; Siliani, S.; Canuti, V.; Rosi, I.; Bertuccioli, M. A kinetic study on extraction and transformation phenomena of phenolic compounds during red wine fermentation. *Int. J. Food Sci. Technol.* **2010**, *45*, 2080–2088. [CrossRef]
6. Cerpa-Calderón, F.K.; Kennedy, J.A. Berry Integrity and Extraction of Skin and Seed Proanthocyanidins during Red Wine Fermentation. *J. Agric. Food Chem.* **2008**, *56*, 9006–9014. [CrossRef]
7. Hernández-Jiménez, A.; Kennedy, J.A.; Bautista-Ortín, A.B.; Gómez-Plaza, E. Effect of Ethanol on Grape Seed Proanthocyanidin Extraction. *Am. J. Enol. Vitic.* **2012**, *63*, 57–61. [CrossRef]
8. Setford, P.C.; Jeffery, D.W.; Grbin, P.R.; Muhlack, R.A. Factors affecting extraction and evolution of phenolic compounds during red wine maceration and the role of process modelling. *Trends Food Sci. Technol.* **2017**, *69*, 106–117. [CrossRef]
9. Marais, J. Effect of Different Wine-Making Techniques on the Composition and Quality of Pinotage Wine. II. Juice/Skin Mixing Practices. *S. Afr. J. Enol. Vitic.* **2003**, *24*. [CrossRef]
10. Togores, J.H. *Tratado de Enología*; Mundi-Prensa: Madrid, España, 2011; ISBN 978-84-8476-531-8.
11. Lowe, E.J.; Oey, A.; Turner, T.M. Gasquet Thermovinification System Perspective after Two Years' Operation. *Am. J. Enol. Vitic.* **1976**, *27*, 130–133.
12. Pezzi, F.; Caprara, C.; Friso, D.; Ranieri, B. Technical and economic evaluation of maceration of red grapes for production everyday wine. *J. Agric. Eng.* **2013**, 323–326. [CrossRef]
13. Morel-Salmi, C.; Souquet, J.-M.; Bes, M.; Cheynier, V. Effect of Flash Release Treatment on Phenolic Extraction and Wine Composition. *J. Agric. Food Chem.* **2006**, *54*, 4270–4276. [CrossRef] [PubMed]
14. Atanacković, M.; Petrović, A.; Jović, S.; Bukarica, L.G.-; Bursać, M.; Cvejić, J. Influence of winemaking techniques on the resveratrol content, total phenolic content and antioxidant potential of red wines. *Food Chem.* **2012**, *131*, 513–518. [CrossRef]
15. Fischer, U.; Strasser, M.; Gutzler, K. Impact of fermentation technology on the phenolic and volatile composition of German red wines. *Int. J. Food Sci. Technol.* **2000**, *35*, 81–94. [CrossRef]
16. Delsart, C.; Ghidossi, R.; Poupot, C.; Cholet, C.; Grimi, N.; Vorobiev, E.; Milisic, V.; Peuchot, M.M. Enhanced Extraction of Phenolic Compounds from Merlot Grapes by Pulsed Electric Field Treatment. *Am. J. Enol. Vitic.* **2012**, *63*, 205–211. [CrossRef]
17. El Darra, N.; Grimi, N.; Maroun, R.G.; Louka, N.; Vorobiev, E. Pulsed electric field, ultrasound, and thermal pretreatments for better phenolic extraction during red fermentation. *Eur. Food Res. Technol.* **2013**, *236*, 47–56. [CrossRef]
18. López, N.; Puértolas, E.; Condón, S.; Álvarez, I.; Raso, J. Application of pulsed electric fields for improving the maceration process during vinification of red wine: Influence of grape variety. *Eur. Food Res. Technol.* **2008**, *227*, 1099. [CrossRef]
19. Leong, S.Y.; Burritt, D.J.; Oey, I. Evaluation of the anthocyanin release and health-promoting properties of Pinot Noir grape juices after pulsed electric fields. *Food Chem.* **2016**, *196*, 833–841. [CrossRef] [PubMed]

20. Rankine, B.C. Heat extraction of color from red grapes of increasing importance. *Wines Vines* **1973**, *54*, 33–36.
21. Girard, B.; Yuksel, D.; Cliff, M.A.; Delaquis, P.; Reynolds, A.G. Vinification effects on the sensory, colour and GC profiles of Pinot noir wines from British Columbia. *Food Res. Int.* **2001**, *34*, 483–499. [CrossRef]
22. Ribéreau-Gayon, P.; Dubourdieu, D.; Donèche, B.; Lonvaud, A. *Handbook of Enology, Volume 1: The Microbiology of Wine and Vinifications*; John Wiley & Sons: Hoboken, NJ, USA, 2006; ISBN 978-0-470-01035-8.
23. Nordestgaard, S. Fermentation: Pre-fermentation heating of red grapes: A useful tool to manage compressed vintages? *Aust. N. Z. Grapegrow. Winemak.* **2017**, *637*, 54–61.
24. Auw, J.M.; Blanco, V.; O'Keefe, S.F.; Sims, C.A. Effect of Processing on the Phenolics and Color of Cabernet Sauvignon, Chambourcin, and Noble Wines and Juices. *Am. J. Enol. Vitic.* **1996**, *47*, 279–286.
25. de Andrade Neves, N.; de Araújo Pantoja, L.; dos Santos, A.S. Thermovinification of grapes from the Cabernet Sauvignon and Pinot Noir varieties using immobilized yeasts. *Eur. Food Res. Technol.* **2014**, *238*, 79–84. [CrossRef]
26. Martinière, P.; Ribéreau-Gayon, J. Modification of the fermentation process by previous heating of the grapes. *Comptes Rendus Hebd. Seances Acad. Sci. Ser. D Sci. Nat.* **1969**, *269*, 925–928.
27. Geffroy, O.; Lopez, R.; Feilhes, C.; Violleau, F.; Kleiber, D.; Favarel, J.-L.; Ferreira, V. Modulating analytical characteristics of thermovinified Carignan musts and the volatile composition of the resulting wines through the heating temperature. *Food Chem.* **2018**, *257*, 7–14. [CrossRef] [PubMed]
28. Niculaua, M.; Tudose-Sandu-Ville, S.; Cotea, V.V.; Luchian, C.E.; Tudose-Sandu-Ville, O.-F. Phenolic Compounds Content in Merlot Wines Obtained through Different Thermomaceration Techniques. *Not. Bot. Horti Agrobot. Cluj-Napoca* **2017**, *45*, 548–552. [CrossRef]
29. Piccardo, D.; González-Neves, G. Extracción de polifenoles y composición de vinos tintos Tannat elaborados por técnicas de maceración prefermentativa. *Agrociencia Urug.* **2013**, *17*, 36–44.
30. Gao, L.; Girard, B.; Mazza, G.; Reynolds, A.G. Changes in Anthocyanins and Color Characteristics of Pinot Noir Wines during Different Vinification Processes. *J. Agric. Food Chem.* **1997**, *45*, 2003–2008. [CrossRef]
31. Geffroy, O.; Lopez, R.; Serrano, E.; Dufourcq, T.; Gracia-Moreno, E.; Cacho, J.; Ferreira, V. Changes in analytical and volatile compositions of red wines induced by pre-fermentation heat treatment of grapes. *Food Chem.* **2015**, *187*, 243–253. [CrossRef]
32. Girard, B.; Kopp, T.G.; Reynolds, A.G.; Cliff, M. Influence of Vinification Treatments on Aroma Constituents and Sensory Descriptors of Pinot noir Wines. *Am. J. Enol. Vitic.* **1997**, *48*, 198.
33. De Boubée, D.R.; Cumsille, A.M.; Pons, M.; Dubourdieu, D. Location of 2-Methoxy-3-isobutylpyrazine in Cabernet Sauvignon Grape Bunches and Its Extractability during Vinification. *Am. J. Enol. Vitic.* **2002**, *53*, 1–5.
34. Markakis, P. Chapter 6-Stability of Anthocyanins in Foods. In *Anthocyanins as Food Colors*; Markakis, P., Ed.; Academic Press: Cambridge, MA, USA, 1982; pp. 163–180. ISBN 978-0-12-472550-8. [CrossRef]
35. Kelebek, H.; Canbas, A.; Selli, S.; Saucier, C.; Jourdes, M.; Glories, Y. Influence of different maceration times on the anthocyanin composition of wines made from Vitis vinifera L. cvs. Boğazkere and Öküzgözü. *J. Food Eng.* **2006**, *77*, 1012–1017. [CrossRef]
36. Morata, A.; Gómez-Cordovés, M.C.; Suberviola, J.; Bartolomé, B.; Colomo, B.; Suárez, J.A. Adsorption of Anthocyanins by Yeast Cell Walls during the Fermentation of Red Wines. *J. Agric. Food Chem.* **2003**, *51*, 4084–4088. [CrossRef] [PubMed]
37. Ageron, D.; Escudier, J.L.; Abbal, P.; Moutounet, M. Prétraitement des raisins par flash détente sous vide poussé. *Rev. Fr. Oenologie* **1995**, *35*, 50–53.
38. Doco, T.; Williams, P.; Cheynier, V. Effect of Flash Release and Pectinolytic Enzyme Treatments on Wine Polysaccharide Composition. *J. Agric. Food Chem.* **2007**, *55*, 6643–6649. [CrossRef] [PubMed]
39. Baggio, P. Flash Extraction–What Can It Do for You? Web page. Available online: https://www.dtpacific.com/dev/wp-content/uploads/2017/03/Art-ASVO-Flash-Bio-Thermo-Extraction-What-can-it-do-for-you.pdf (accessed on 1 April 2018).
40. Vernhet, A.; Bes, M.; Bouissou, D.; Carrillo, S.; Brillouet, J.-M. Characterization of suspended solids in thermo-treated red musts. *J. Int. Sci. Vigne Vin* **2016**, *50*, 9. [CrossRef]
41. Escudier, J.L.; Bes, M.; Morel-Salmi, C.; Micolajczak, M.; Sanson, A. Vinification en rouge: Macérations post-fermentaires, macérations carboniques, flash-détente sous vide. *Rev. Fr. OEnol.* **2006**, *216*, 11–19.
42. Kotséridis, Y.; Escudier, J.L.; Moutounet, M. Flash détente et qualité des vins. Progr. *Agric. Vitic.* **2002**, *20*, 438–442.

43. Razungles, A. Extraction technologies and wine quality. In *Managing Wine Quality*; Elsevier: Amsterdam, The Netherlands, 2010; pp. 589–630. ISBN 978-1-84569-798-3. [CrossRef]
44. Besnard, É.; Laffargue, F.; Relhié, F.; Fro, H.; Alibert, V. Influence de l'itinéraire de vinification après Flash-détente dans l'élaboration d'une nouvelle gamme de vins du Lot. Web page. Available online: http://www.vignevin-occitanie.com/wp-content/uploads/2018/08/flash-detente-lot-malbec.pdf (accessed on 1 April 2019).
45. Toepfl, S.; Mathys, A.; Heinz, V.; Knorr, D. Review: Potential of High Hydrostatic Pressure and Pulsed Electric Fields for Energy Efficient and Environmentally Friendly Food Processing. *Food Rev. Int.* **2006**, *22*, 405–423. [CrossRef]
46. Chemat, F.; Rombaut, N.; Meullemiestre, A.; Turk, M.; Perino, S.; Fabiano-Tixier, A.-S.; Abert-Vian, M. Review of Green Food Processing techniques. Preservation, transformation, and extraction. *Innov. Food Sci. Emerg. Technol.* **2017**, *41*, 357–377. [CrossRef]
47. Tsong, T.Y. Electroporation of Cell Membranes. In *Electroporation and Electrofusion in Cell Biology*; Neumann, E., Sowers, A.E., Jordan, C.A., Eds.; Springer: Boston, MA, USA, 1989; pp. 149–163. ISBN 978-1-4899-2528-2. [CrossRef]
48. Weaver, J.C.; Chizmadzhev, Y.A. Theory of electroporation: A review. *Bioelectrochem. Bioenerg.* **1996**, *41*, 135–160. [CrossRef]
49. Toepfl, S. Pulsed electric field food processing –industrial equipment design and commercial applications. *Stewart Postharvest Rev.* **2012**, *8*, 1–7. [CrossRef]
50. Puértolas, E.; Luengo, E.; Álvarez, I.; Raso, J. Improving Mass Transfer to Soften Tissues by Pulsed Electric Fields: Fundamentals and Applications. *Annu. Rev. Food Sci. Technol.* **2012**, *3*, 263–282. [CrossRef] [PubMed]
51. Luengo, E.; Franco, E.; Ballesteros, F.; Álvarez, I.; Raso, J. Winery Trial on Application of Pulsed Electric Fields for Improving Vinification of Garnacha Grapes. *Food Bioprocess Technol.* **2014**, *7*, 1457–1464. [CrossRef]
52. Toepfl, S. Pulsed Electric Field food treatment-scale up from lab to industrial scale. *Procedia Food Sci.* **2011**, *1*, 776–779. [CrossRef]
53. Puértolas, E.; López, N.; Condón, S.; Álvarez, I.; Raso, J. Potential applications of PEF to improve red wine quality. *Trends Food Sci. Technol.* **2010**, *21*, 247–255. [CrossRef]
54. Cholet, C.; Delsart, C.; Petrel, M.; Gontier, E.; Grimi, N.; L'Hyvernay, A.; Ghidossi, R.; Vorobiev, E.; Mietton-Peuchot, M.; Gény, L. Structural and Biochemical Changes Induced by Pulsed Electric Field Treatments on Cabernet Sauvignon Grape Berry Skins: Impact on Cell Wall Total Tannins and Polysaccharides. *J. Agric. Food Chem.* **2014**, *62*, 2925–2934. [CrossRef]
55. Saldaña, G.; Cebrián, G.; Abenoza, M.; Sánchez-Gimeno, C.; Álvarez, I.; Raso, J. Assessing the efficacy of PEF treatments for improving polyphenol extraction during red wine vinifications. *Innov. Food Sci. Emerg. Technol.* **2017**, *39*, 179–187. [CrossRef]
56. Ricci, A.; Parpinello, G.P.; Versari, A. Recent Advances and Applications of Pulsed Electric Fields (PEF) to Improve Polyphenol Extraction and Color Release during Red Winemaking. *Beverages* **2018**, *4*, 18. [CrossRef]
57. Puértolas, E.; Hernández-Orte, P.; Saldaña, G.; Álvarez, I.; Raso, J. Improvement of winemaking process using pulsed electric fields at pilot-plant scale. Evolution of chromatic parameters and phenolic content of Cabernet Sauvignon red wines. *Food Res. Int.* **2010**, *43*, 761–766. [CrossRef]
58. Puértolas, E.; Saldaña, G.; Condón, S.; Álvarez, I.; Raso, J. Evolution of polyphenolic compounds in red wine from Cabernet Sauvignon grapes processed by pulsed electric fields during aging in bottle. *Food Chem.* **2010**, *119*, 1063–1070. [CrossRef]
59. López-Alfaro, I.; González-Arenzana, L.; López, N.; Santamaría, P.; López, R.; Garde-Cerdán, T. Pulsed electric field treatment enhanced stilbene content in Graciano, Tempranillo and Grenache grape varieties. *Food Chem.* **2013**, *141*, 3759–3765. [CrossRef]
60. Garde-Cerdán, T.; González-Arenzana, L.; López, N.; López, R.; Santamaría, P.; López-Alfaro, I. Effect of different pulsed electric field treatments on the volatile composition of Graciano, Tempranillo and Grenache grape varieties. *Innov. Food Sci. Emerg. Technol.* **2013**, *20*, 91–99. [CrossRef]
61. Cravotto, G.; Cintas, P. Power ultrasound in organic synthesis: Moving cavitational chemistry from academia to innovative and large-scale applications. *Chem. Soc. Rev.* **2006**, *35*, 180–196. [CrossRef]
62. Chemat, F.; Rombaut, N.; Sicaire, A.-G.; Meullemiestre, A.; Fabiano-Tixier, A.-S.; Abert-Vian, M. Ultrasound assisted extraction of food and natural products. Mechanisms, techniques, combinations, protocols and applications. A review. *Ultrason. Sonochem.* **2017**, *34*, 540–560. [CrossRef]

63. González-Centeno, M.R.; Knoerzer, K.; Sabarez, H.; Simal, S.; Rosselló, C.; Femenia, A. Effect of acoustic frequency and power density on the aqueous ultrasonic-assisted extraction of grape pomace (Vitis vinifera L.)—A response surface approach. *Ultrason. Sonochem.* **2014**, *21*, 2176–2184. [CrossRef]
64. Celotti, E.; Ferraretto, P. Studies for the ultrasound application in winemaking for a low impact enology. In Proceedings of the 39th World Congress of Vine and Wine, Bento Gonçalves, Brazil, 23–28 October 2016; p. 132.
65. Zhang, Q.-A.; Shen, Y.; Fan, X.-H.; García Martín, J.F. Preliminary study of the effect of ultrasound on physicochemical properties of red wine. *CyTA-J. Food* **2016**, *14*, 55–64. [CrossRef]
66. Bautista-Ortín, A.B.; Jiménez-Martínez, M.D.; Jurado, R.; Iniesta, J.A.; Terrades, S.; Andrés, A.; Gómez-Plaza, E. Application of high-power ultrasounds during red wine vinification. *Int. J. Food Sci. Technol.* **2017**, *52*, 1314–1323. [CrossRef]
67. Del Fresno, J.M.; Morata, A.; Escott, C.; Loira, I.; Cuerda, R.; Suárez-Lepe, J. Sonication of Yeast Biomasses to Improve the Ageing on Lees Technique in Red Wines. *Molecules* **2019**, *24*, 635. [CrossRef]
68. Singleton, V.L.; Draper, D.E. Ultrasonic Treatment with Gas Purging as a Quick Aging Treatment for Wine. *Am. J. Enol. Vitic.* **1963**, *14*, 23–35.

Article

Effect of the Atmospheric Pressure Cold Plasma Treatment on Tempranillo Red Wine Quality in Batch and Flow Systems

Elisa Sainz-García [1], Isabel López-Alfaro [2], Rodolfo Múgica-Vidal [1], Rosa López [2], Rocío Escribano-Viana [2], Javier Portu [2], Fernando Alba-Elías [1] and Lucía González-Arenzana [2,*]

1 Departamento de Ingeniería Mecánica, Universidad de La Rioja, Calle San José de Calasanz, 31, 26004 Logroño, Spain

2 Instituto de Ciencias de la Vid y del Vino, ICVV (Gobierno de La Rioja, Universidad de La Rioja and CSIC), Finca La Grajera, Ctra. Burgos, Km. 6, 26007 Logroño, Spain

* Correspondence: lucia.gonzalez@icvv.es; Tel.: +34-94-189-4980

Received: 27 May 2019; Accepted: 18 July 2019; Published: 5 August 2019

Abstract: The demand for chemical-free beverages is posing a challenge to the wine industry to provide safe and healthy products with low concentrations of chemical preservatives. The development of new technologies, such as Atmospheric Pressure Cold Plasma (APCP), offers the wine industry the opportunity to contribute to this continuous improvement. The purpose of this research is to evaluate the effect of Argon APCP treatment, applied in both batch and flow systems, on Tempranillo red wine quality. Batch treatments of 100 mL were applied with two powers (60 and 90 W) at four periods (1, 3, 5, and 10 min). For flowing devices, 750 mL of wine with a flow of 1.2 and 2.4 L/min were treated at 60 and 90 W for 25 min and was sampled every 5 min. Treatments in batch resulted in wines with greater color intensity, lower tonality, and higher content in total phenolic compounds and anthocyanins, so that they were favorable for wine quality. Among the batch treatments, the one with the lowest power was the most favorable. Flow continuous treatments, despite being more appropriate to implement in wineries, neither led to significant improvements in the chromatic and phenolic wine properties nor caused wine spoilage.

Keywords: atmospheric pressure cold plasma; continuous flow; batch; Argon; red wine; color

1. Introduction

Current consumers are looking for natural, safe, and healthy beverages with low concentrations of chemical preservatives, thereby maintaining their healthy properties [1,2]. The cutting-edge technologies for food processing offers the wine industry the opportunity to contribute to this improvement. Thus, technology based on the application of Atmospheric Pressure Cold Plasma (APCP) is a very attractive innovation tool for the food industry.

Plasma is a state of matter, similar to gas, in which some particles are ionized. The APCP uses different gases such as air, nitrogen, argon, helium, etc., that are applied directly or indirectly (through a liquid medium) to process and disinfect materials. In the application of the APCP, a large number and diversity of highly energetic reactive species are generated. This activates physical and chemical processes difficult to achieve in ordinary chemical environments. In fact, plasma is a source of UV photons, charged particles (positive and negative ions), free radicals, atoms, and molecules excited or not, etc., with a high antimicrobial capacity [3].

To date, among the various physical and chemical food decontamination techniques evaluated, APCP has demonstrated a high efficiency in the reduction of microbial contaminants in different foods and beverages [4–6]. In contrast, there are few studies evaluating how APCP affects the quality of the treated foods and beverages, despite being determinant to the consumer election.

The color is one of the most important sensory attributes as it is normally the first feature perceived by the wine consumer, consequently having a great influence on the final wine quality perception [7]. Wine color does not only depend on the initial grape composition, but also on the different techniques applied in the cellar and the numerous biochemical reactions that occurred during the winemaking process [8–10]. Phenolic compounds are responsible for the organoleptic characteristics of wine, such as color and flavor. Moreover, some of these compounds have been related to antioxidant and free radical-scavenging properties [3,11] that may play a role in human health, including a highly probable protection against cardiovascular diseases and cancer [12–14].

Some results have shown a decrease in total phenolics and total flavonoids in grape juice and an increase in total flavonols after high voltage atmospheric cold plasma treatments [15]. Other authors have studied the stability of phenolic compounds in several fruit juices, and even in wines [16,17], concluding that plasma treatments have an impact that is mainly dependent on the equipment and processing parameters.

Moreover, most of the plasma experiments have been conducted in static conditions. This circumstance makes it even more difficult to implement in real wineries, so that APCP application in continuous flow might be a great advance for winemakers. The main goal of this study was to examine the influence of continuous flow APCP in chromatic characteristics and the phenolic compounds content of red wines, compared to batch APCP treatments.

2. Materials and Methods

2.1. Wine and APCP Treatments

This study was conducted with a young Tempranillo red wine from the ICVV experimental winery, sampled just after the spontaneous MLF of the harvest 2017, with 12% alcoholic strength and a pH of 3.65.

Six different treatments were carried out by a non-thermal atmospheric pressure plasma jet system (PlasmaSpot®, VITO, Boeretang, Belgium). This system consists of a plasma torch that operates at atmospheric pressure, with two cylindrical electrodes in coaxial arrangement that are separated by a dielectric barrier of Al_2O_3. A flow of Argon gas (40 slm) was supplied in all cases. Any excess gas in the system was evacuated through an exhaust by a fan. The frequency of the generator was fixed at 68 kHz for the entire process.

The plasma was applied in two ways: In batch and continuous flow systems (Figure 1). Thus, 100 mL of wine was treated in batch with plasma running at the combination of the following processing parameters: Power at 60 and 90 W and treatment time of 1, 3, 5, and 10 min. Moreover, 750 mL of wine was treated in continuous flow for 1.2 and 2.4 L/min at powers of 60 and 90 W, and treatment time of 25 min, sampling every 5 min. The treatments in batch were named with B (batch) and with LF (low flow) or HF (high flow), and the power (60 or 90) (Table 1). The six different treatments (B60, B90, LF60, HF60, LF90, and HF90) were carried out independently with biological triplicates ($n = 3$) and the application was conducted during six days over two weeks. Before treatments, all samples were at room temperature.

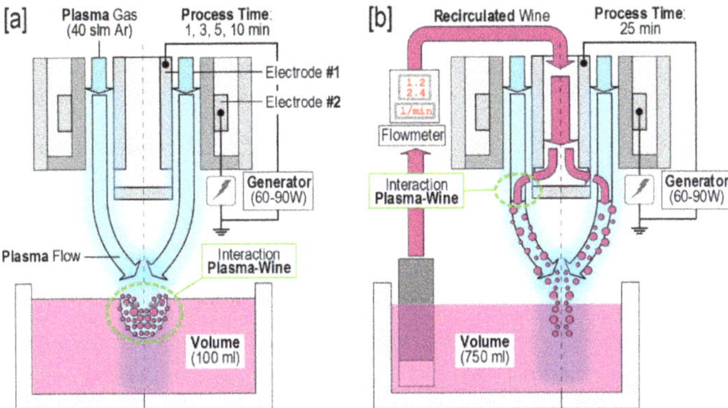

Figure 1. Schematic representation of (**a**) batch Atmospheric Pressure Cold Plasma (APCP) treatments and of (**b**) continuous APCP treatments.

Table 1. Argon APCP treatments in batch (B) and low flow (LF) and high flow (HF) systems (F) characteristics.

Treatment	Power (W)	Batch	Wine Flow (mL/h)	Volume (mL)	Time (min)
B60	60	Yes	No	100	0,1,3,5,10
B90	90	Yes	No	100	0,1,3,5,10
LF60	60	No	1.2	750	0,10,15,20,25
HF60	60	No	2.4	750	0,10,15,20,25
LF90	90	No	1.2	750	0,10,15,20,25
HF90	90	No	2.4	750	0,10,15,20,25

2.2. *Analysis of Physical and Color Parameters*

Before and after each treatment, every sample was analyzed regarding the physical parameters of temperature, pH, and conductivity with a multi-meter of temperature, pH, and conductivity (multisensor 5048, HACH, Madrid, Spain). Color intensity (CI) and hue were measured according to the European Community Official Methods protocols [18]. Total phenolic compounds in mg/L of gallic acid (TP) were determined by the Folin-Ciocalteau method by the Miura-One enzymatic auto analyzer (TDI S.L., Barcelona, Spain).

2.3. *Analysis of Anthocyanins and Vitisins by HPLC*

In addition, the samples were analyzed in terms of individual anthocyanins and vitisins by HPLC. Non-acylated and acylated anthocyanins and vitisins were analysed using an Agilent 1260 Infinity chromatograph, equipped with a diode array detector (DAD, Agilent, Santa Clara, CA, USA). The procedure followed was described by Portu et al. [19]. They used a Licrospher® 100 RP-18 reversed-phase column (250 × 4.0 mm; 5 μm packing; Agilent, Santa Clara, CA, USA) with pre-column Licrospher® 100 RP-18 (4 × 4 mm; 5 μm packing; Agilent, Santa Clara, CA, USA). The column temperature was set at 40 °C, the flow rate was established at 0.630 mL/min, and the injection volume was 10 μL. Eluents were (A) acetonitrile/water/formic acid (3:88.5:8.5, *v*/*v*/*v*), and (B) acetonitrile/water/formic acid (50:41.5:8.5, *v*/*v*/*v*). The linear solvent gradient was: 0 min, 6% B; 15 min, 30% B; 30 min, 50% B; 35 min, 60% B; 38 min, 60% B; 46 min, and 6% B.

Anthocyanins were identified according to the retention times of the available pure compounds and the UV–Vis data obtained from authentic standards and/or published in previous studies on β-glucosidase activity. Anthocyanins were quantified at 520 nm as malvidin-3-*O*-glucoside (Extrasynthèse, Genay, France). Concentrations were expressed as milligrams per litter of wine (mg/L).

The data corresponds to the average of the analyses of three samples ($n = 3$). The total anthocyanins consisted of the sum of the individualized anthocyanins.

2.4. Statistical Analysis

The analytical parameters measured for each of the samples were statistically analyzed with the SPSS software (IBM® SPSS Statistic version 23, Armonk, NY, USA). Analyses of the variance (ANOVA) were assessed. The significant differences between mean values were determined by Tukey's HSD test and differences were considered as significant when the p value was below 0.05.

3. Results and Discussion

This study analyzed the effects of Argon APCP batch and flow systems applied to a red wine and it was focused on its color and phenolic properties. For that purpose, the treatments were applied with two different powers, in batch and flowing, and during several minutes. Variability between starting wines was mainly due to the fact that treatments were performed in six days of two consecutive weeks. The initial wine suffered some type of evolution during those days, which made it that times zero or control samples had different physicochemical parameters like the pH. This made us consider each treatment as totally independent.

3.1. Impact of APCP Treatments on Physical and Color Parameters

Results of the average physical parameters measured in wines before and after the six Argon APCP treatments are shown in Table 2. The temperature of the control and treated samples hardly varied during the application of most of the treatments. In batch systems, the temperature did not change significantly after 10 min of treatment, the same result was observed in low flow systems after 25 min. The temperature was only significantly increased with time in the treatments linked to high flow systems, although it varied only from 19.7 °C to 20.4 °C when applying 60 W and from 20.7 °C to 21.8 °C when applying 90 W. This result demonstrated the cold character of APCP treatments despite the applied energy [20]. The pH parameter measures the hydrogen ion concentration of a solution. A decrease of water pH after APCP treatments was demonstrated [21]. In contrast, inconclusive results about pH variation after APCP have been observed with other products [22]. In our study, only after the batch treatment of low power (B60) applied to wine from 1 to 5 min did the pH significantly increase to reach around 4 units, which might trigger the microbial spoilage of wine. On another point, the conductivity means that the facility of a liquid media might flow an electric discharge. It has been observed that plasma activated water conductivity is higher than the conductivity found in non-treated water [21]. In this research, wine conductivity was significantly higher than the control after 3 or 5 min of batch APCP treatments. In contrast, conductivity was reduced after flowing treatments, although this reduction was only significant after 20 min of treatment HF90. This reduction was also observed by Pankaj et al. [15] after APCP treatment of 80 kV for 4 min in grape juice. In general, a drop in the conductivity of wines is related to a loss in their tartaric stability [23].

Table 2. Average physical parameters assessed for wine with different Argon APCP treatments and the standard deviation of data ($n = 3$).

Treatment	Time (min)	T (°C)	pH	C (μS)
B60	0	21.4 ± 0.0	3.77 ± 0.00 ab	1750 ± 0 a
	1	21.3 ± 0.0	3.93 ± 0.19 b	1760 ± 17 a
	3	21.1 ± 0.0	3.93 ± 0.10 b	1787 ± 6 ab
	5	21.0 ± 0.0	3.89 ± 0.04 b	1830 ± 30 b
	10	19.9 ± 0.0	3.61 ± 0.00 a	1903 ± 25 c
B90	0	21.4 ± 0.0	3.77 ± 0.00	1750 ± 0 b
	1	21.3 ± 0.0	3.85 ± 0.27	1730 ± 0 a
	3	21.2 ± 0.0	3.60 ± 0.03	1770 ± 10 c
	5	21.0 ± 0.0	3.66 ± 0.07	1847 ± 12 d
	10	22.0 ± 0.0	3.69 ± 0.15	1923 ± 6 e
LF60	0	20.0 ± 0.1	3.69 ± 0.02	1873 ± 15
	5	19.9 ± 0.1	3.70 ± 0.02	1867 ± 12
	10	19.8 ± 0.1	3.71 ± 0.03	1850 ± 10
	15	19.7 ± 0.1	3.62 ± 0.02	1793 ± 57
	20	19.6 ± 0.2	3.69 ± 0.09	1553 ± 453
	25	19.7 ± 0.4	3.72 ± 0.07	1460 ± 624
HF60	0	19.7 ± 0.3 ab	3.47 ± 0.30	1590 ± 243
	5	19.5 ± 0.2 a	3.38 ± 0.30	1583 ± 240
	10	19.7 ± 0.1 a	3.47 ± 0.23	1560 ± 217
	15	19.9 ± 0.0 ab	3.49 ± 0.25	1517 ± 168
	20	20.2 ± 0.1 bc	3.53 ± 0.20	1450 ± 280
	25	20.4 ± 0.1 c	3.58 ± 0.13	1213 ± 68
LF90	0	20.0 ± 0.3 ab	3.60 ± 0.06	1460 ± 52
	5	19.7 ± 0.5 a	3.55 ± 0.11	1460 ± 10
	10	20.1 ± 0.6 ab	3.56 ± 0.07	1457 ± 12
	15	20.2 ± 0.3 ab	3.46 ± 0.13	1447 ± 6
	20	20.7 ± 0.3 ab	3.47 ± 0.14	1367 ± 42
	25	20.9 ± 0.3 b	3.52 ± 0.09	1343 ± 93
HF90	0	20.7 ± 0.2 b	3.70 ± 0.05	1755 ± 5 b
	5	20.3 ± 0.2 a	3.70 ± 0.07	1760 ± 10 b
	10	20.7 ± 0.2 b	3.74 ± 0.08	1740 ± 20 b
	15	21.1 ± 0.1 c	3.79 ± 0.13	1737 ± 15 b
	20	21.5 ± 0.1 d	3.81 ± 0.14	1617 ± 47 a
	25	21.8 ± 0.1 d	3.81 ± 0.15	1560 ± 20 a

Nomenclature abbreviations: T. temperature; C. conductivity. Different letters mean significant differences between the samples of the same treatment ($p \leq 0.05$). No letters mean no significant differences.

The color intensity (CI) is an index of the amount of color of a wine. It significantly increased after batch treatments (Table 3). This increase was of approximately two units for both 60 and 90 W, which is, overall, positive for red wine quality [16]. However, the CI increase after flow systems was lower (between 0.2 and 0.6 units after 25 min of treatment) and had statistical significance only for LF60, HF90, and LF90 treatments (Table 3). The lowest hue of a wine means a positive wine evolution, which was observed after 10 min of batch treatments. Flow APCP treatments also resulted in a tonality reduction, although it was only significant in the treatment of the greatest flow and power.

Table 3. Average color parameters assessed for wine with different Argon APCP treatment and the standard deviation of data (n = 3).

Treatment	Time (min)	CI	Hue	TP (mg/L Gallic Acid)
B60	0	6.50 ± 0.00 a	0.599 ± 0.00 d	1067 ± 71 a
	1	8.30 ± 0.05 ab	0.579 ± 0.00 c	1120 ± 8 ab
	3	8.37 ± 0.01 ab	0.571 ± 0.00 b	1132 ± 14 ab
	5	8.42 ± 0.05 ab	0.567 ± 0.00 a	1140 ± 11 ab
	10	8.47 ± 0.08 b	0.563 ± 0.00 a	1170 ± 24 b
B90	0	6.50 ± 0.00 a	0.599 ± 0.00 d	1067 ± 71
	1	7.36 ± 0.02 b	0.592 ± 0.00 cd	1132 ± 26
	3	7.66 ± 0.04 c	0.587 ± 0.00 bc	1145 ± 30
	5	7.94 ± 0.02 d	0.581 ± 0.00 b	1155 ± 22
	10	8.24 ± 0.17 e	0.564 ± 0.01a	1194 ± 26
LF60	0	6.72 ± 0.15 a	0.590 ± 0.01	1044 ± 29
	5	6.94 ± 0.05 ab	0.592 ± 0.01	1049 ± 55
	10	6.98 ± 0.16 ab	0.587 ± 0.01	1063 ± 33
	15	7.05 ± 0.12 ab	0.585 ± 0.01	1081 ± 45
	20	7.08 ± 0.10 ab	0.581 ± 0.01	1077 ± 16
	25	7.19 ± 0.15 b	0.581 ± 0.01	1066 ± 12
HF60	0	7.47 ± 0.56	0.593 ± 0.01	1062 ± 59
	5	7.54 ± 0.54	0.592 ± 0.01	1058 ± 66
	10	7.56 ± 0.51	0.589 ± 0.01	1062 ± 56
	15	7.64 ± 0.54	0.589 ± 0.01	1084 ± 60
	20	7.72 ± 0.51	0.586 ± 0.00	1067 ± 13
	25	7.78 ± 0.55	0.583 ± 0.00	1105 ± 41
LF90	0	7.82 ± 0.06	0.588 ± 0.00	1087 ± 81
	5	7.89 ± 0.10	0.590 ± 0.00	1070 ± 70
	10	7.91 ± 0.12	0.593 ± 0.00	1078 ± 71
	15	7.97 ± 0.02	0.587 ± 0.00	1089 ± 62
	20	8.00 ± 0.06	0.587 ± 0.00	1041 ± 8
	25	8.02 ± 0.06	0.585 ± 0.00	1052 ± 19
HF90	0	6.49 ± 0.03 a	0.608 ± 0.00 c	1087 ± 81
	5	6.66 ± 0.05 b	0.604 ± 0.01 bc	1097 ± 48
	10	6.75 ± 0.02 bc	0.599 ± 0.00 abc	1103 ± 51
	15	6.84 ± 0.06 cd	0.599 ± 0.00 abc	1108 ± 57
	20	6.95 ± 0.04 d	0.595 ± 0.00 ab	1087 ± 54
	25	7.09 ± 0.04 e	0.593 ± 0.00 a	1101 ± 42

Nomenclature abbreviations: CI Color index; TP Total Polyphenols. Different letters mean significant differences between the samples of the same treatment ($p \leq 0.05$). No letters mean no significant differences.

The average total phenolic (TP) compounds, determined by the reaction with the Folin reagent, was significantly higher after batch APCP treatments, varying from 1067 to 1170 with 60 W and from 1067 to 1194 with 90 W (Table 3). In this way, Herceg et al. [24] reported an increase in TP in pomegranate juice after Argon plasma treatment. This index is based on the capacity of the phenolics to react with oxidant agents so that it is a total determination of phenolic compounds, but it also expresses the contribution of these compounds to the antioxidant activity of the sample, so its increase is positive for wine quality. However, flowing APCP treatments did not result in a significant modification of TP. Even in the bibliography, contradictory results are found, for instance Lukić et al. [25] observed a reduction of TP in a Cabernet Sauvignon red wine, which could be due to the possible degradation of these compounds by the plasma mechanism.

3.2. *Impact of APCP Treatments on Anthocyanin and Vitisins Contents*

The anthocyanins free monomers are the main responsible for the color of young red wines. They were analyzed individually for every sample. The HPLC analyses identified five non-acylated anthocyanins (Table 4), including 3-o-glucosides (3-glc) of delphinidin, cyanidin, petunidin, peonidin, and malvidin and 11 acylated anthocyanins (Table 5), including acetyl glucosides (3-acglc) of delphinidin, petunidin, peonidin, and malvidin; trans-p-coumaroyl glucosides (3-cmglc) of delphinidin, cyanidin, petunidin, peonidin, and cis and trans malvidin and caffeoyl glucoside (cfglc) of malvidin.

Table 4. Average non-acylated anthocyanins (mg/L) wine with different Argon APCP treatments and the standard deviation of data ($n = 3$).

Treatment	Time (min)	Dp-3-glc	Cn-3-glc	Pt-3-glc	Pn-3-glc	Mv-3-glc
B60	0	13.2 ± 0.5 a	2.97 ± 0.03 b	20.3 ± 0.5 a	7.64 ± 0.32	139 ± 5 a
	1	14.1 ± 0.1 b	2.89 ± 0.02 ab	22.3 ± 0.2 b	7.90 ± 0.13	149 ± 2 b
	3	14.0 ± 0.2 ab	2.91 ± 0.05 ab	21.8 ± 0.1 b	7.82 ± 0.06	148 ± 1 b
	5	14.2 ± 0.2 b	2.86 ± 0.02 a	22.1 ± 0.3 b	8.05 ± 0.05	150 ± 1 b
	10	14.2 ± 0.4 b	2.92 ± 0.05 ab	22.2 ± 0.4 b	7.94 ± 0.26	149 ± 3 b
B90	0	13.2 ± 0.5 a	2.97 ± 0.03	20.3 ± 0.5 a	7.64 ± 0.32	139 ± 5
	1	14.3 ± 0.4 b	2.98 ± 0.07	22.2 ± 0.5 c	8.09 ± 0.34	147 ± 2
	3	13.7 ± 0.1 ab	2.94 ± 0.04	21.3 ± 0.2 abc	7.60 ± 0.09	144 ± 1
	5	13.4 ± 0.5 ab	3.01 ± 0.02	20.9 ± 0.5 bc	7.53 ± 0.20	142 ± 4
	10	13.9 ± 0.3 ab	2.96 ± 0.04	21.8 ± 0.4 ab	7.83 ± 0.19	146 ± 3
LF60	0	10.4 ± 0.6	2.90 ± 0.08	16.8 ± 0.8	6.55 ± 0.33 b	113 ± 6 b
	5	10.3 ± 0.5	2.90 ± 0.05	16.5 ± 0.4	6.47 ± 0.10 ab	112 ± 2 ab
	15	10.1 ± 0.5	2.96 ± 0.11	16.0 ± 0.5	6.34 ± 0.19 ab	110 ± 3 ab
	25	9.5 ± 0.4	2.96 ± 0.08	15.4 ± 0.5	5.95 ± 0.21 a	102 ± 4 a
HF60	0	10.8 ± 1.0	2.91 ± 0.08	17.3 ± 1.3	6.63 ± 0.39	118 ± 9
	5	11.2 ± 0.5	2.90 ± 0.11	17.9 ± 1.0	6.85 ± 0.40	121 ± 6
	15	11.2 ± 0.5	2.97 ± 0.07	17.6 ± 0.9	6.68 ± 0.36	119 ± 6
	25	10.6 ± 0.8	3.00 ± 0.10	17.2 ± 1.5	6.32 ± 0.44	114 ± 10
LF90	0	10.1 ± 0.1	2.84 ± 0.05	16.2 ± 0.2	6.31 ± 0.16	109 ± 2
	5	10.9 ± 0.6	2.87 ± 0.04	17.6 ± 0.9	6.61 ± 0.19	118 ± 5
	15	10.9 ± 0.6	2.91 ± 0.04	17.5 ± 0.4	6.62 ± 0.15	118 ± 5
	25	10.6 ± 0.6	2.89 ± 0.11	16.9 ± 0.8	6.42 ± 0.16	114 ± 5
LF90	0	13.2 ± 0.5	2.97 ± 0.03	20.3 ± 0.5	7.64 ± 0.32	139 ± 5
	5	13.3 ± 0.3	3.00 ± 0.01	20.9 ± 0.5	7.70 ± 0.15	142 ± 3
	15	12.9 ± 0.3	3.04 ± 0.08	20.5 ± 0.8	7.50 ± 0.27	138 ± 5
	25	12.6 ± 0.3	2.99 ± 0.02	20.0 ± 0.5	7.27 ± 0.18	151 ± 25

Nomenclature abbreviations: Dp. delphinidin; Cn. cyanidin; Pt. petunidin; Pn. peonidin; Mv. malvidin; glc. glucoside. All parameters are listed with their standard deviation ($n = 3$). Different letters mean significant differences between the sample of the same treatment ($p \leq 0.01$). No letters mean no significant differences.

Non-acylated anthocyanins represented around 70%, with malvidin-3-o-glucoside being the majority. Between non-acylated anthocyanins, malvidin derivatives were also found to be the predominant anthocyanin type, while coumaroylated anthocyanins were the major acylated form, which is in accordance with previous studies with the Tempranillo grape variety [25]. Cyanidin -3- acetylglucoside was not detected in any case. The most important variations for non-acylated anthocyanins were obtained for batch treatments (Table 4). Thus, APCP static treatment, applied with a power of 60 W, led to a significant increase of delphinidin, petunidin, and malvidin-3-glucosides; the same treatment applied with a power of 90 W enhanced delphinidin and petunidin-3 glucosides. In both cases, the positive increase for the wine quality of these compounds occurred from the first minute of treatment and remained practically constant. This increase could be related to the observed increase in CI exposed previously. As Table 4 shows, there were hardly any variations for the flowing APCP treatments. Only peonidin and malvidin-3-glc decreased significantly after 25 min of the treatment with the lowest flow and power.

Table 5. Average acylated anthocyanin content (mg/L) (acetylated, coumarylated and caffeoylated) in wine with different Argon APCP treatments.

Treatment	Time (min)	Acetylated				Coumaroylated						Caffeoylated
		Dp-3-acglc	Pt-3-acglc	Pn-3-acglc	Mv-3-acglc	Cn-3-cmglc	Dp-3-cmglc	Pt-3-cmglc	Pn-3-cmglc	Mv-3-Cis-cmglc	Mv-3-Trans-cmglc	Mv-3-cfGlc
B60	0	2.64 ± 0.05	3.28 ± 0.05 a	2.34 ± 0.02	7.87 ± 0.25 a	2.32 ± 0.02	3.32 ± 0.05 a	3.26 ± 0.10 a	3.73 ± 0.09 a	3.57 ± 0.11 ab	10.54 ± 0.64 a	2.69 ± 0.03 a
	1	2.61 ± 0.04	3.36 ± 0.03 ab	2.35 ± 0.01	8.47 ± 0.09 b	2.37 ± 0.06	3.64 ± 0.14 b	3.46 ± 0.10 b	3.75 ± 0.10 ab	3.60 ± 0.06 ab	11.45 ± 0.17 ab	2.71 ± 0.01 ab
	3	2.62 ± 0.01	3.40 ± 0.01 b	2.33 ± 0.02	8.37 ± 0.06 b	2.35 ± 0.02	3.48 ± 0.03 ab	3.47 ± 0.04 b	3.97 ± 0.09 b	3.57 ± 0.03 ab	11.80 ± 0.17 b	2.73 ± 0.03 ab
	5	2.63 ± 0.05	3.46 ± 0.04 b	2.35 ± 0.01	8.38 ± 0.06 b	2.35 ± 0.03	3.54 ± 0.08 ab	3.36 ± 0.07 ab	3.76 ± 0.09 ab	3.46 ± 0.14 a	11.23 ± 0.08 ab	2.75 ± 0.06 ab
	10	2.67 ± 0.02	3.44 ± 0.07 b	2.33 ± 0.02	8.27 ± 0.12 b	2.32 ± 0.00	3.52 ± 0.12 ab	3.38 ± 0.02 ab	3.86 ± 0.04 ab	3.72 ± 0.07 b	11.50 ± 0.40 b	2.77 ± 0.02 b
B90	0	2.64 ± 0.05	3.28 ± 0.05 a	2.34 ± 0.02	7.87 ± 0.25 a	2.32 ± 0.02	3.32 ± 0.05 a	3.26 ± 0.10	3.73 ± 0.09	3.57 ± 0.11	10.54 ± 0.64 a	2.69 ± 0.03
	1	2.63 ± 0.03	3.47 ± 0.07 b	2.34 ± 0.03	8.56 ± 0.18 b	2.30 ± 0.01	3.54 ± 0.10 b	3.41 ± 0.05	3.87 ± 0.15	3.61 ± 0.09	11.45 ± 0.73 b	2.71 ± 0.06
	3	2.65 ± 0.03	3.34 ± 0.03 ab	2.32 ± 0.02	8.19 ± 0.01 ab	2.31 ± 0.02	3.42 ± 0.06 ab	3.25 ± 0.04	3.80 ± 0.16	3.63 ± 0.03	10.91 ± 0.46 ab	2.69 ± 0.01
	5	2.71 ± 0.07	3.28 ± 0.05 ab	2.31 ± 0.03	8.11 ± 0.22 ab	2.33 ± 0.01	3.40 ± 0.07 ab	3.25 ± 0.04	3.72 ± 0.09	3.68 ± 0.14	10.74 ± 0.23 ab	2.71 ± 0.01
	10	2.66 ± 0.04	3.43 ± 0.09 b	2.32 ± 0.01	8.21 ± 0.07 ab	2.32 ± 0.01	3.43 ± 0.05 ab	3.30 ± 0.07	3.72 ± 0.04	3.60 ± 0.10	10.81 ± 0.26 ab	2.71 ± 0.01
LF60	0	2.66 ± 0.08	3.11 ± 0.17 b	2.30 ± 0.06	6.71 ± 0.28	2.30 ± 0.01	2.92 ± 0.04	2.92 ± 0.06	3.28 ± 0.21	3.61 ± 0.17	8.30 ± 0.73	2.55 ± 0.06
	5	2.61 ± 0.02	2.97 ± 0.14 ab	2.24 ± 0.02	6.58 ± 0.13	2.26 ± 0.01	2.96 ± 0.02	2.90 ± 0.06	3.28 ± 0.05	3.62 ± 0.09	8.13 ± 0.24	2.57 ± 0.04
	15	2.64 ± 0.09	2.98 ± 0.03 ab	2.31 ± 0.06	6.64 ± 0.19	2.25 ± 0.04	2.90 ± 0.07	2.90 ± 0.04	3.25 ± 0.21	3.65 ± 0.06	8.13 ± 0.69	2.51 ± 0.02
	25	2.63 ± 0.05	2.93 ± 0.02 a	2.28 ± 0.03	6.41 ± 0.15	2.26 ± 0.03	2.94 ± 0.02	2.79 ± 0.02	3.27 ± 0.20	3.74 ± 0.10	7.72 ± 0.51	2.52 ± 0.05
HF60	0	2.63 ± 0.03	3.05 ± 0.10	2.29 ± 0.01	6.82 ± 0.46	2.24 ± 0.01	3.03 ± 0.08	2.93 ± 0.08	3.26 ± 0.09	3.65 ± 0.02	8.48 ± 0.72	2.52 ± 0.04
	5	2.65 ± 0.03	3.11 ± 0.09	2.28 ± 0.02	7.04 ± 0.19	2.26 ± 0.01	3.04 ± 0.10	3.00 ± 0.10	3.45 ± 0.11	3.62 ± 0.10	9.12 ± 0.81	2.55 ± 0.04
	15	2.63 ± 0.06	3.04 ± 0.07	2.30 ± 0.01	6.95 ± 0.19	2.26 ± 0.01	3.13 ± 0.06	2.95 ± 0.07	3.21 ± 0.09	3.61 ± 0.12	8.44 ± 0.40	2.60 ± 0.06
	25	2.63 ± 0.03	3.03 ± 0.13	2.27 ± 0.01	6.73 ± 0.42	2.24 ± 0.02	2.97 ± 0.07	2.90 ± 0.07	3.27 ± 0.18	3.68 ± 0.06	8.21 ± 0.96	2.53 ± 0.03
LF90	0	2.65 ± 0.05	3.03 ± 0.10	2.27 ± 0.03	6.65 ± 0.22	2.26 ± 0.03	2.94 ± 0.02	2.81 ± 0.04	3.39 ± 0.14	3.66 ± 0.19	8.22 ± 0.34	2.50 ± 0.02
	5	2.66 ± 0.07	3.09 ± 0.09	2.30 ± 0.01	6.71 ± 0.15	2.27 ± 0.03	3.04 ± 0.02	2.91 ± 0.10	3.36 ± 0.04	3.53 ± 0.14	8.68 ± 0.23	2.52 ± 0.02
	15	2.63 ± 0.05	3.08 ± 0.04	2.28 ± 0.05	6.79 ± 0.24	2.29 ± 0.04	3.10 ± 0.15	2.91 ± 0.07	3.44 ± 0.10	3.67 ± 0.14	8.81 ± 0.26	2.58 ± 0.08
	25	2.66 ± 0.04	3.06 ± 0.10	2.29 ± 0.03	6.93 ± 0.24	2.26 ± 0.01	3.07 ± 0.07	2.95 ± 0.07	3.37 ± 0.17	3.61 ± 0.14	8.63 ± 0.54	2.60 ± 0.11
LF90	0	2.64 ± 0.05	3.28 ± 0.05	2.34 ± 0.02	7.87 ± 0.25	2.32 ± 0.02	3.32 ± 0.05	3.26 ± 0.10	3.73 ± 0.09	3.57 ± 0.11	10.54 ± 0.64	2.69 ± 0.03
	5	2.63 ± 0.05	3.37 ± 0.07	2.34 ± 0.03	8.24 ± 0.25	2.32 ± 0.02	3.36 ± 0.05	3.26 ± 0.12	3.69 ± 0.11	3.61 ± 0.10	10.55 ± 0.37	2.70 ± 0.03
	15	2.62 ± 0.03	3.31 ± 0.07	2.32 ± 0.02	8.03 ± 0.35	2.31 ± 0.01	3.36 ± 0.04	3.20 ± 0.07	3.60 ± 0.05	3.68 ± 0.08	10.36 ± 0.4	2.70 ± 0.05
	25	2.61 ± 0.01	3.23 ± 0.08	2.33 ± 0.01	7.92 ± 0.17	2.32 ± 0.02	3.28 ± 0.07	3.13 ± 0.03	3.60 ± 0.08	3.70 ± 0.06	9.89 ± 0.37	2.64 ± 0.04

Nomenclature abbreviations: Dp. delphinidin; Cn. cyanidin; Pt. petunidin; Pn. peonidin; Mv. malvidin; glc. glucoside; acglc. acetylglucoside; cmglc. *trans-p*-coumaroylglucoside; cfglc. caffeoylglucoside. All parameters are listed with their standard deviation ($n = 3$). Different letters mean significant differences between the samples of the same treatment ($p \leq 0.01$). No letters mean no significant differences.

Similar to anthocyanins non-acylated, the most important variations for the acylated ones were obtained for batch treatments (Table 5). Thus, for B60, a significant increase was observed for petunidin and malvidin-3-acglc, for five of the six coumaroylated glucosides, and for malvidin-3-cfglc. The treatment time from which the concentration of these compounds increased varied according to each of them (between 1 min and 10 min). For B90, a lower number of anthocyanins increased their concentration significantly, including petunidin and malvidin-3-acglc, and delphinidin and malvidin-3-trans-cmglc. In this case, as happened with non-acylated anthocyanins, the increase only occurred after one minute of treatment and remained practically constant. As can be observed in Table 5, dynamic treatments did not produce variations in these compounds, with the exception of petudine-3-acglc, which decreased significantly after 25 min of low flow treatment and 60 watts of power. Elez et al. [26] applied plasma to sour cherry Marasca and they found a higher concentration of anthocyanins compared to untreated juice for short treatments (3 min). However, results obtained by Lukić [16] indicated a decrease in the composition of free anthocyanins in a Cabernet Sauvignon red wine. This decrease became more sever with the treatment duration and the frequency of the batch treatment. This could be explained by the degradation of these compounds by the plasma mechanisms.

Due to the difficulty to evaluate every individual compound in the samples, their total contents were studied (Figure 2). It was demonstrated that treatment of B60 led to an increase of the total acylated, non-acylated, and the total anthocyanins. Applying the highest power of APCP in the batch did not cause this impact. This was because the total anthocyanins content only increased after 1 min of treatment with statistical significance, but this effect disappeared with longer treatments. Being anthocyanins, which are the compounds mainly responsible for red wine color, the batch treatment with the lowest power was the most favorable for the color of the wine. On the other hand, none of the dynamic treatments did not significantly modify the total content of acylated anthocyanins, nor non-acylated, and nor the total anthocyanins content.

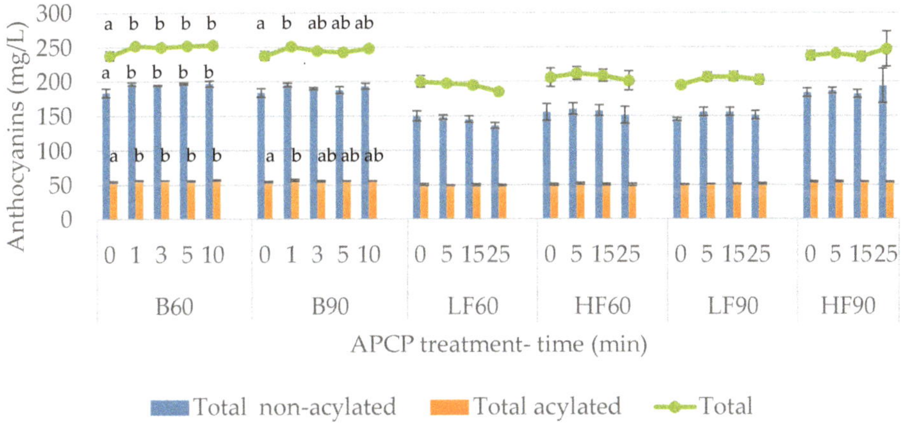

Figure 2. Total anthocyanins content (mg/l) of non-acetylated and acetylated anthocyanins in wine after different Argon-APCP treatments applied different times (0, 5, 15, and 25 min). Error bars represent the standard deviation ($n = 3$). Different letters mean significant differences between samples ($p \leq 0.01$). No letters mean no significant differences.

In Figure 3, results of vitisins A and B of every sample are shown. Some anthocyanin derived pigments from malvidin-3-glucoside, such as vitisins A and B are of interest because of their stability in the conditions common in red wines [27]. Treatment of B60 was the only one that caused a significant increase in vitisin A content after 10 min of treatment, but the lowest content was described 3 min after APCP treatments. However, this effect was not observed in vitisin B content with this same treatment. In this case, flow treatment HF90 caused a significant increase of vitisin B 25 min after treatment. The rest of the treatments did not modify the concentration of these compounds.

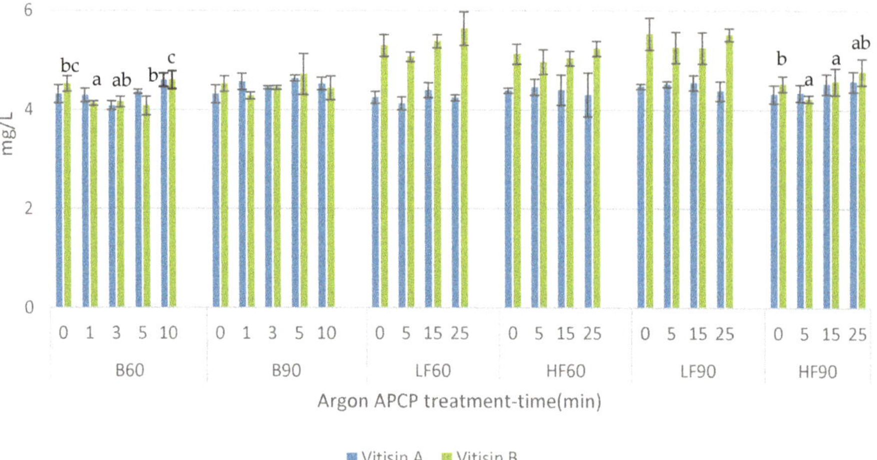

Figure 3. Vitisin A and B content (mg/L) in wine after different Argon–APCP treatments applied different times (0, 1, 3, 5, 10, 15, and 25 min). Error bars represent the standard deviation ($n = 3$). Different letters mean significant differences between samples ($p \leq 0.01$). No letters mean no significant differences.

4. Conclusions

Important variations were not observed in the physical parameters of Tempranillo young wine after Argon APCP treatments, except in the pH increase after 5 min of low power batch treatment. Nevertheless, chromatic properties and phenolic compounds content depended on the treatment conditions. Overall, treatments in batch caused a more favorable impact on wine quality, since they provided greater color intensity, lower tonality, and higher content in total phenolic compounds and anthocyanins of wine. Among the batch treatments, the one with lower power was the most positive, which could mean that the highest energies are not necessarily linked to better effects of the technique under the color properties of red wine. The utilization of flowing argon APCP treatments in wines, in spite of being the most appropriate system to be implemented in wineries, did not lead to significant improvements in the chromatic properties and phenolic compounds content in the wine, but they were not unfavorable. Further research and investigations should be carried out to improve the results of APCP flowing systems.

Author Contributions: F.A.-E. and I.L.-A. designed the experiment. E.S.-G. and R.M.-V. were in charge of PLASMA equipment and application of treatments. R.E.-V., J.P., L.G.-A. and I.L.-A. carried out the analytical assessments. R.L., I.L.-A. and L.G.-A. wrote and reviewed the manuscript. L.G.-A. submitted the manuscript and acted as correspondence author.

Funding: This work was funded by the Government of La Rioja project R-11-18 that could be co-financed by the European Regional Development Fund, granted to the Autonomous Community of La Rioja, within the ERDF Operational Program.

Conflicts of Interest: The authors declare no conflict of interest.

References

1. Costanigro, M.; Appleby, C.; Menke, S.D. The Wine Headache: Consumer Perceptions of Sulfites and Willingness to Pay for Non-Sulfited Wines. *Food Qual. Prefer.* **2014**, *31*, 81–89. [CrossRef]
2. D'Amico, M.; Di Vita, G.; Monaco, L. Exploring Environmental Consciousness and Consumer Preferences for Organic Wines without Sulfites. *J. Clean. Prod.* **2016**, *120*, 64–71. [CrossRef]

3. Arnous, A.; Makris, D.P.; Kefalas, P. Effect of Principal Polyphenolic Components in Relation to Antioxidant Characteristics of Aged Red Wines. *J. Agric. Food Chem.* **2001**, *49*, 5736–5742. [CrossRef] [PubMed]
4. Ziuzina, D.; Boehm, D.; Patil, S.; Cullen, P.J.; Bourke, P. Cold Plasma Inactivation of Bacterial Biofilms and Reduction of Quorum Sensing Regulated Virulence Factors. *PLoS ONE* **2015**, *10*, e0138209. [CrossRef]
5. Misra, N.; Schlüter, O.; Cullen, P. *Cold Plasma in Food and Agriculture: Fundamentals and Applications*; Academic Press Elsevier: Amsterdam, The Nederland, 2016.
6. Kim, H.-J.; Yong, H.I.; Park, S.; Kim, K.; Choe, W.; Jo, C. Microbial Safety and Quality Attributes of Milk Following Treatment with Atmospheric Pressure Encapsulated Dielectric Barrier Discharge Plasma. *Food Control* **2015**, *47*, 451–456. [CrossRef]
7. Parpinello, G.P.; Versari, A.; Chinnici, F.; Galassi, S. Relationship among Sensory Descriptors, Consumer Preference and Color Parameters of Italian Novello Red Wines. *Food Res. Int.* **2009**, *42*, 1389–1395. [CrossRef]
8. Barsotti, L.; Dumay, E.; Mu, T.H.; Fernandez Diaz, M.D.; Cheftel, J.C. Effects of High Voltage Electric Pulses on Protein-Based Food Constituents and Structures. *Trends Food Sci. Technol.* **2001**, *12*, 136–144. [CrossRef]
9. García-Beneytez, E.; Revilla, E.; Cabello, F. Anthocyanin Pattern of Several Red Grape Cultivars and Wines Made from Them. *Eur. Food Res. Technol.* **2002**, *215*, 32–37. [CrossRef]
10. Sacchi, K.L.; Bisson, L.F.; Adams, D.O. A Review of the Effect of Winemaking Techniques on Phenolic Extraction in Red Wines. *Am. J. Enol. Vitic.* **2005**, *56*, 197–206.
11. Minussi, R.C.; Rossi, M.; Bologna, L.; Cordi, L.; Rotilio, D.; Pastore, G.M.; Durán, N. Phenolic Compounds and Total Antioxidant Potential of Commercial Wines. *Food Chem.* **2003**, *82*, 409–416. [CrossRef]
12. Briviba, K.; Abrahamse, S.L.; Pool-Zobel, B.L.; Rechkemmer, G. Neurotensin-and EGF-Induced Metabolic Activation of Colon Carcinoma Cells Is Diminished by Dietary Flavonoid Cyanidin but Not by Its Glycosides. *Nutr. Cancer* **2001**, *41*, 172–179. [CrossRef]
13. Negro, C.; Tommasi, L.; Miceli, A. Phenolic Compounds and Antioxidant Activity from Red Grape Marc Extracts. *Bioresour. Technol.* **2003**, *87*, 41–44. [CrossRef]
14. Pérez-Magariño, S.; González-San José, M.L. Polyphenols and Colour Variability of Red Wines Made from Grapes Harvested at Different Ripeness Grade. *Food Chem.* **2006**, *96*, 197–208. [CrossRef]
15. Pankaj, S.K.; Wan, Z.; Colonna, W.; Keener, K.M. Effect of High Voltage Atmospheric Cold Plasma on White Grape Juice Quality. *J. Sci. Food Agric.* **2017**, *97*, 4016–4021. [CrossRef] [PubMed]
16. Lukić, K.; Vukušić, T.; Tomašević, M.; Ćurko, N.; Gracin, L.; Kovačević Ganić, K. The Impact of High Voltage Electrical Discharge Plasma on the Chromatic Characteristics and Phenolic Composition of Red and White Wines. *Innov. Food Sci. Emerg. Technol.* **2017**, *53*, 70–77. [CrossRef]
17. Bursać Kovačević, D.; Putnik, P.; Dragović-Uzelac, V.; Pedisić, S.; Režek Jambrak, A.; Herceg, Z. Effects of Cold Atmospheric Gas Phase Plasma on Anthocyanins and Color in Pomegranate Juice. *Food Chem.* **2016**, *190*, 317–323. [CrossRef] [PubMed]
18. Commission Regulation. Commission Regulation (EC) No 606/2009 of 10 July 2009 Laying down Certain Detailed Rules for Implementing Council Regulation (EC) No 479/2008 as Regards the Categories of Grapevine Products, Oenological Practices and the Applicable Restrictions. *Off. J. Eur. Union L* **2009**, *193*, 1–59.
19. Portu, J.; López, R.; Baroja, E.; Santamaría, P.; Garde-Cerdán, T. Improvement of Grape and Wine Phenolic Content by Foliar Application to Grapevine of Three Different Elicitors: Methyl Jasmonate, Chitosan, and Yeast Extract. *Food Chem.* **2016**, *201*, 213–221. [CrossRef]
20. Mandal, R.; Singh, A.; Singh, A.P. Recent Developments in Cold Plasma Decontamination Technology in the Food Industry. *Trends Food Sci. Technol.* **2018**, *80*, 93–103. [CrossRef]
21. Thirumdas, R.; Kothakota, A.; Annapure, U.; Siliveru, K.; Blundell, R.; Gatt, R.; Valdramidis, V.P. Plasma Activated Water (PAW): Chemistry, Physico-Chemical Properties, Applications in Food and Agriculture. *Trends Food Sci. Technol.* **2018**, *77*, 21–31. [CrossRef]
22. Coutinho, N.M.; Silveira, M.R.; Rocha, R.S.; Moraes, J.; Ferreira, M.V.S.; Pimentel, T.C.; Freitas, M.Q.; Silva, M.C.; Raices, R.S.L.; Ranadheera, C.S.; et al. Cold Plasma Processing of Milk and Dairy Products. *Trends Food Sci. Technol.* **2018**, *74*, 56–68. [CrossRef]
23. Bosso, A.; Motta, S.; Petrozziello, M.; Guaita, M.; Asproudi, A.; Panero, L. Validation of a Rapid Conductimetric Test for the Measurement of Wine Tartaric Stability. *Food Chem.* **2016**, *212*, 821–827. [CrossRef]
24. Herceg, Z.; Kovačević, D.B.; Kljusurić, J.G.; Jambrak, A.R.; Zorić, Z.; Dragović-Uzelac, V. Gas Phase Plasma Impact on Phenolic Compounds in Pomegranate Juice. *Food Chem.* **2016**, *190*, 665–672. [CrossRef]

25. Portu, J.; López-Alfaro, I.; Gómez-Alonso, S.; López, R.; Garde-Cerdán, T. Changes on Grape Phenolic Composition Induced by Grapevine Foliar Applications of Phenylalanine and Urea. *Food Chem.* **2015**, *180*, 171–180. [CrossRef]
26. Elez Garofulić, I.; Režek Jambrak, A.; Milošević, S.; Dragović-Uzelac, V.; Zorić, Z.; Herceg, Z. The Effect of Gas Phase Plasma Treatment on the Anthocyanin and Phenolic Acid Content of Sour Cherry Marasca (Prunus Cerasus Var. Marasca) Juice. *LWT-Food Sci. Technol.* **2015**, *62*, 894–900. [CrossRef]
27. Morata, A.; Calderón, F.; González, M.C.; Gómez-Cordovés, M.C.; Suárez, J.A. Formation of the Highly Stable Pyranoanthocyanins (Vitisins A and B) in Red Wines by the Addition of Pyruvic Acid and Acetaldehyde. *Food Chem.* **2007**, *100*, 1144–1152. [CrossRef]

Review

Pulsed Light: Challenges of a Non-Thermal Sanitation Technology in the Winemaking Industry

Aitana Santamera, Carlos Escott, Iris Loira, Juan Manuel del Fresno, Carmen González and Antonio Morata *

EnotecUPM, Chemistry and Food Technology Department, Universidad Politécnica de Madrid, Avenida Puerta de Hierro 2, 28040 Madrid, Spain; a.santamera@alumnos.upm.es (A.S.); carlos.escott@gmail.com (C.E.); iris.loira@upm.es (I.L.); juanmanuel.delfresno@upm.es (J.M.d.F.); carmen.gchamorro@upm.es (C.G.)

* Correspondence: antonio.morata@upm.es; Tel.: +34-910-671-127

Received: 6 June 2020; Accepted: 10 July 2020; Published: 14 July 2020

Abstract: Pulsed light is an emerging non-thermal technology viable for foodstuff sanitation. The sanitation is produced through the use of high energy pulses during ultra-short periods of time (ns to μs). The pulsed light induces irreversible damages at the DNA level with the formation of pyrimidine dimers, but also produces photo-thermal and photo-physical effects on the microbial membranes that lead to a reduction in the microbial populations. The reduction caused in the microbial populations can reach several fold, up to 4 log CFU/mL decrement. A slight increase of 3 to 4 °C in temperature is observed in treated food; nonetheless, this increase does not modify either the nutritional properties of the product or its sensory profile. The advantages of using pulsed light could be used to a greater extent in the winemaking industry. Experimental trials have shown a positive effect of reducing native yeast and bacteria in grapes to populations below 1–2 log CFU/mL. In this way, pulsed light, a non-thermal technology currently available for the sanitation of foodstuffs, is an alternative for the reduction in native microbiota and the later control of the fermentative process in winemaking. This certainly would allow the use of fermentation biotechnologies such as the use of non-*Saccharomyces* yeasts in mixed and sequential fermentations to preserve freshness in wines through the production of aroma volatile compounds and organic acids, and the production of wines with less utilization of SO_2 in accordance with the consumers' demand in the market.

Keywords: antimicrobial; food technology; non-*Saccharomyces*; enzymatic activity; wine quality

---

1. Introduction

Pulsed light is composed of white light comprising the visible light spectrum and fractions of the ultraviolet and near infrared invisible light spectra [1] that can be obtained from different sources, with silica fibers doped with ytterbium ions (Yb^{3+}) being one of them [2]. This material is able to produce pulses with ultrashort durations (picoseconds and femtoseconds) and very high energy. Another source, commonly used in commercial equipment, is inert gases flash lamps filled with xenon or krypton [3]. Either way, the effects produced by a sequence of high intensity pulses has been tested in many different industrial fields.

Most of the initial experimental trials on the use of this technology were performed in the second half of the 20th century, in particular during the last three decades. The diversity of applications involving the use of pulsed light covers, to mention some, the thermal process of localized surfaces in semiconductors [4] without affecting the temperature of the overall device, or the possibility of sintering copper nanoparticles by replacing conventional thermal sintering, not suitable for the production of conductive lines, with a reactive sintering method for printed electronics with such a material [5]. In other distant field, the dermatology and cosmetic industries have also thoroughly evaluated the use of a variation of pulsed light technology with cut-off filters to select prescribed wavelengths [6] in treatments able to improve the aspect of the skin or to work as photo-rejuvenation [7] or promote collagen formation in upper dermal skin layers [8]; to work effectively on vascular facial lesions produced in patients with rosacea [9] or patients with facial hemangiomas [10]; to improve the state of stretch marks [11], and even to remove corporal hair in a long-term effective epilation process which is safe for people to use [12].

The effects that pulsed light have on biological structures have led to the use of this technology in other scopes. It has been observed that the energy released during the ultrashort emission treatments may affect protein structures and cellular membranes or even promote nucleic acid destruction and dimer formation [13]. These advantages have proven to work as alternatives for the reduction in pathogens from food matrices [14,15] that have an impact, not only on the shelf life and quality of many foodstuffs intended for human consumption, but also on the health and safety demanded in these products.

This review provides up to date accessible information regarding this emerging non-thermal technology towards its utilization in the food industry, in particular its feasibility on the industrial usage and scale-up for grape and must sanitation in the winemaking industry.

2. Pulsed Light Treatment Mechanism

The energy involved in the pulsed light (PL) technology comprises the production of photons released by atoms when they are excited and then relaxed to a lower energy state. The photons could be emitted from a continuous light source or, in the case of PL, in pulses, the mechanism of which increases the energy, as the emission is produced in a short time [3]. The energy generated is stored in capacitors able to keep this energy over short periods of time—fractions of a second—and then release the energy over the surfaces to be treated [16]. The energy emitted in pulses is increased with this arrangement and it can be estimated to be several times the energy of the sun received by the surface of the planet at sea level [17].

One of the reasons why PL has antimicrobial properties is because the light emitted covers a wide range of the electromagnetic spectrum, especially the UV radiation. The PL energy covers from high frequency wavelengths in the UV spectrum (≈200 nm) to wavelengths in the near infrared spectrum (2500 nm) (Figure 1). The radiation corresponding to the UV spectrum includes all three wavelength ranges: long, medium and short, or UV-A, UV-B and UV-C, respectively [3].

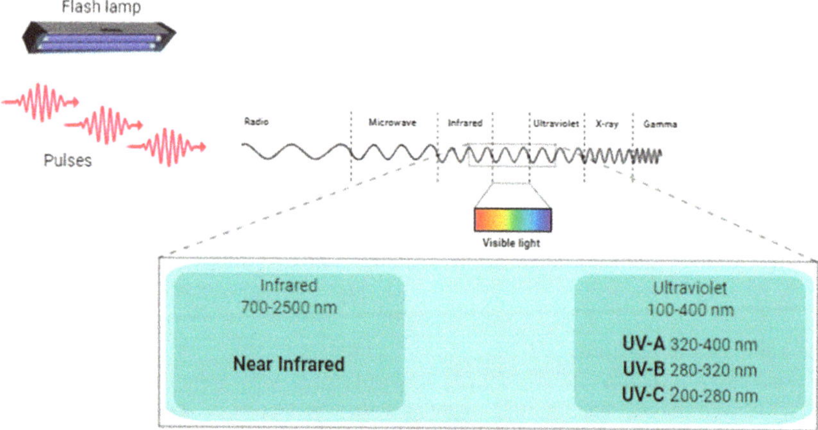

Figure 1. Range of the electromagnetic spectrum covered in the pulsed light (PL) irradiation.

The UV-C light fraction from the PL radiation is mainly responsible for the inactivation of microorganisms found on the surface of the foodstuff, although UV-B and UV-A fractions can also contribute to microbial inactivation [18]. The UV-C light corresponds to the range of the electromagnetic spectrum between 200 and 280 nm [19]. The photochemical effect of the UV-C radiation (254–260 nm) is responsible for the formation of pyrimidine dimers, new covalent bonds previously non-existing in the polynucleotide chain, that inhibits the formation of new DNA chains needed for the cell replication [16,19]. The dimer formation occurs between adjacent thymine bases or between thymine and cytosine bases and inhibits the DNA replication. Although the principal consequence of the photochemical effect is the formation of dimers, there is also evidence on the occurrence of single-strand breaks and double-strand breaks of DNA's structure. Besides this photochemical effect, there might also be photo-thermal and photo-physical damage in the biological structures of yeast, bacteria and viruses when using PL. These damages include changes in membrane permeability, depolarization of the cell membrane, ion flow variations and localized heating [3] (Figure 2). At the same time, photo-induced degradation of phenolic compounds has been reported when PL treatments above 3.8 J/cm^2 are used in liquid media [20]. This suggests a potential reduction in secondary plant metabolites such as phenolic acids and flavonoids in uncontrolled treatments, with a subsequent reduction in the antioxidant capacity of the foodstuff.

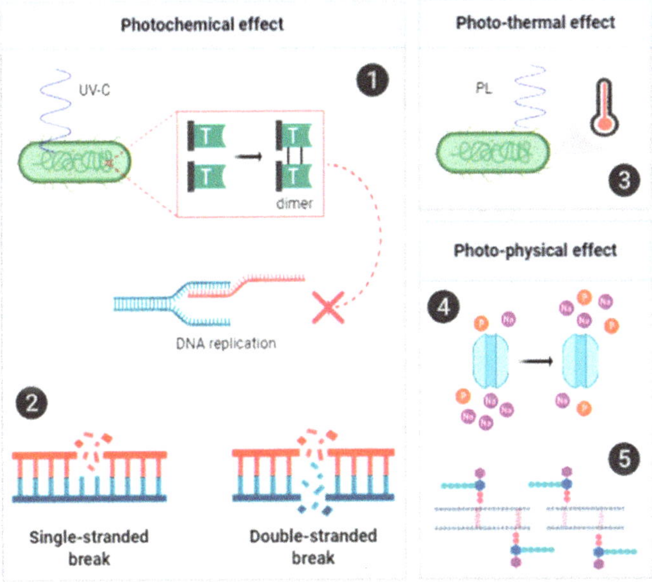

Figure 2. Photochemical, photo-thermal and photo-physical effects of UV-C radiation on microorganisms. (**1**) formation of dimers from adjacent thymines and inhibition of DNA replication, (**2**) single- and double-stranded breaks of DNA chains, (**3**) locally increased temperature, (**4**) ion flow modification and (**5**) membrane permeation.

In PL technology, for an efficient antimicrobial activity, it is critical to consider the number of pulses, the exposure time, and the dose or fluence that the product is receiving during the treatment, which is expressed as cumulative energy input (J/cm^2) [21]. Dose-wise, there are several factors that influence this parameter since UV photons are less energetic than other photons; therefore, the distance between the flash lamp and the surface to be treated, the shading produce by the geometry of the sample and the micro-shading produced by the roughness of the surface and the discharge intensity are factors than can limit the effectiveness of treatments considering limited amount of pulses. The intensity of one emitted pulse can reach from 0.1 up to 50 J/cm^2 at the surface with up to 20 flashes per second [1]. Such a range would make it possible to use these pulses to inactivate microorganisms naturally found on foodstuff surfaces considering the FDA cumulative recommendation of 12 J/cm^2 for food treatments [22].

Even though there is a photo-thermal effect on biological structures produced locally, with temperature rising up to 130 °C, causing the rapid heating of microbial membranes [23], the overall thermal effect on particular food matrices can be considered negligible. PL together with other emerging technologies such high pressure processing (HPP) or high hydrostatic pressure (HHP), pulsed electric field (PEF) and e-beam irradiation are considered non-thermal technologies able to preserve food nutritional and organoleptic properties contrary to what is observed when using conventional thermal technologies [24]. Experimental trials using a pulsed light sterilization laboratory unit with two xenon flash lamps of 254 mm length and maximum energy of 6 kVA have shown an increase in temperature between 3 and 4 °C (Figure 3). The increase in temperature observed on the surface of the berries would not imply an alteration of the sensory profile, nor would it change the nutritional properties either.

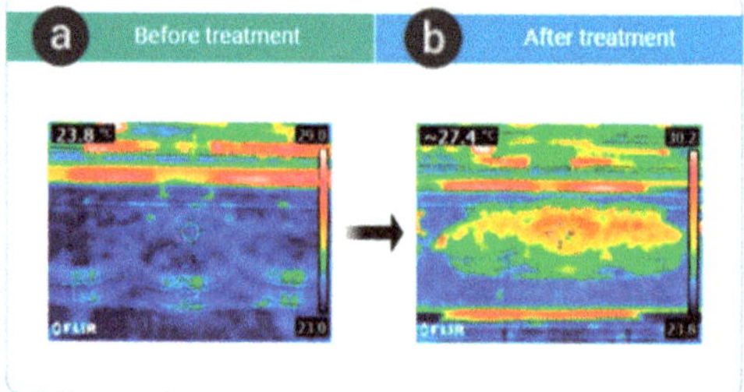

Figure 3. Temperature increase on the surface of table grapes after 5-pulses treatment. (**a**) temperature on the surface before pulses, and (**b**) temperature after pulses treatment.

Lastly, the use of PL may also imply having more energy efficient processes [25], as well as lower operative costs and minimal environmental impact [3], since there are no emissions of organic volatiles or suspended particles related to this emerging technology.

3. Food Spoilage Microorganisms and Pathogens Elimination

PL has been evaluated for the reduction in a wide variety of microorganisms located on the most outermost layer of non-processed products and others multiplied in crossed contamination during the transformation into ready-to-eat foodstuff. The control of microbial populations with PL includes reducing or eliminating spoilage yeasts, molds and bacteria [26,27] and assuring the elimination of pathogens including *Escherichia coli* [28], *Salmonella enterica* [29] and *Listeria monocytogenes* [30,31], the virulent species of which are responsible for diseases such as gastroenteritis, salmonellosis and listeriosis. Foodborne pathogens are responsible for millions of illnesses around the world every year and thus, the number of deaths related to foodborne diseases and the health-related costs can be significant [32].

The effectiveness of PL in the reduction in pathogens and food spoilage microorganisms has been tested in several different foodstuff matrices in the last decades. The use of PL in fruits and vegetables, either on raw products or after any kind of food processing (slicing, cutting, etc.), is shown in Table 1, whilst Table 2 shows data regarding the microbial reduction observed in meat products. Information regarding the use of this sanitation technology in grapes, must and wine is still scarce and limited to a low number of publications.

Table 1. Microbial reduction levels for fruits and vegetables after pulsed light treatment.

Food Product	Microorganism	Processing Conditions	Reduction (log10 CFU/mL)	Reference
Raspberry	*Escherichia coli O157:H7* *Salmonella* Total viable count Yeast and moulds	Total fluence (J/cm^2): 28.2 Peak power (J/cm^2/pulse): 1.27 Exposure time (s): 30 Distance from the lamp (cm): 13	3.9 4.5 1.5 1.6	[33]
Tomato fruit	Microflora in skin and peduncle *Saccharomyces cerevisiae*	Total fluence (J/cm^2): 4 (microflora) and 2.2 (Cerevisiae) Pulse width (µs): 250 Discharge voltage (V): 2500	1 2.3	[34]
Avocado	Aerobic mesophilic microorganisms	Total fluence (J/cm^2): 14 Peak power (J/cm^2/pulse): 0.4 Pulse width (µs): 300 Distance from the lamp (cm): 5	1.2	[35]
Fresh-cut tomatoes	*Psychrophilic bacteria* Moulds and yeasts	Total fluence (J/cm^2): 4, 6 and 8 Peak power (J/cm^2/pulse): 0.4 Pulse width (µs): 300 Stored for 20 days at 4 °C	Up to 1.8 Up to 0.5	[36]
Fresh-cut tomato	*Listeria innocua* *Escherichia coli*	Total fluence (J/cm^2): 4, 6 and 8 Peak power (J/cm^2/pulse): 0.4 Pulse width (µs): 300 Distance above and below from the lamp (cm): 8.5 Stored for 20 days at 4 °C	0.9 1.4	[28]
Spinach leaves	*Listeria innocua* *Escherichia coli*	Total fluence (J/cm^2): 0.8 Peak power (J/cm^2/pulse): 0.4 Pulse width (µs): 300 Distance from the lamp (cm): 8.5	1.85 1.72	[26]
	Mesophilic aerobic bacteria *Psychrotrophic bacteria* *Coliforms* *Listeria ssp* *Yeast and moulds*	Total fluence (J/cm^2): 4 Pulse width (µs): 300 Distance from the lamp (cm): 8.5	0.5–2.2	

Table 1. *Cont.*

Food Product	Microorganism	Processing Conditions	Reduction (log10 CFU/mL)	Reference
Fresh-cut apple slices	Mesophilic and psychrophilic *aerobic bacteria* *Moulds and yeasts*	Total fluence (J/cm^2): 16 Peak power (J/cm^2/pulse): 0.4 Pulse width (μs): 300 Distance above and below from the lamp (cm): 8.5	1.55 2.3	[27]
Fresh-cut apple slices	Total viable counts *Lactobacillus brevis* *Listeria monocytogenes*	Number of pulses: 9 Peak power (J/cm^2/pulse): 1.75 Pulse width (μs): 500 Distance from the lamp (cm): 1	1.0 3.0 2.7	[30]
Cherry tomato	*Salmonella enterica*	Total fluence (J/cm^2): 31.5 Peak power (J/cm^2/pulse): 0.35 Pulse width (μs): 500 Exposure time (s): 30 Distance from the lamp (cm): 14	2.3	[29]
Fresh-cut cucumber slices	*Escherichia coli* ATCC 26	Total fluence (J/cm^2): 12 Peak power (J/cm^2/pulse): 0.43 Pulse width (μs): 360 Exposure time (s): 12.4 Distance from the lamp (cm): 10.8	2.8	[37]
Fresh-cut avocado, watermelon and mushrooms	*Escherichia coli* *Listeria innocua*	Total fluence (J/cm^2): 12 Peak power (J/cm^2/pulse): 0.4 Pulse width (μs): 300 Distance from the lamp (cm): 8.5	2.58, 2.88 and 2.97 2.25, 2.17 and 3.62	[38]
Strawberries (S) and blueberries (B)	*Murine norovirus* (MNV-1) *Escherichia coli* *Salmonella*	Total fluence (J/cm^2): 22.5 Exposure time (s): 24 Number of pulses: 16 Distance from the lamp (cm): 16	S: 0.9 B: 3.8 S: 1.9 B: 5.7 S: 2.1 B: 4.2	[39]
Blueberries	*Salmonella*	Total fluence (J/cm^2): 6 Peak power (J/cm^2/pulse): 0.066 Pulse width (μs): 360 Exposure time (s): 30	0.9 spot inoculation 0.6 dip inoculation	[40]

Table 1. *Cont.*

Food Product	Microorganism	Processing Conditions	Reduction (log10 CFU/mL)	Reference
Fresh-cut lettuce	*Salmonella enteritidis* *Escherichia coli* *Staphylococcus aureus* *Listeria monocytogenes*	Total fluence (J/cm^2): 16.8 Peak power (J/cm^2/pulse): 0.33 Pulse width (μs): 300 Exposure time (s): 25 Distance from the lamp (cm): 9	5.40 5.08 6.56 4.00	[31]
	Total bacteria count Yeast and moulds	Total fluence (J/cm^2): 4–16.8 Peak power (J/cm^2/pulse): 0.33 Pulse width (μμs): 300 Exposure time (s): 6–25 Distance from the lamp (cm): 9	2.73 1.14	
Raspberries	*Salmonella Newport* *Escherichia coli*	Total fluence (J/cm^2): 14.3 Peak power (J/cm^2/pulse): 1.27 Exposure time (s): 15	3.4 3.3	[41]
Green onion	*Escherichia coli*	Total fluence (J/cm^2): 5 Peak power (J/cm^2/pulse): 1.27 Exposure time (s): 5	Spot inoculation: stems 4.8 and leaves 4.1 Dip inoculation: stems 0.2 and leaves 0.6	[42]

Table 2. Microbial reduction levels for meat after pulsed light treatment.

Food Product	Microorganism	Treatment	Reduction (log10 CFU/mL)	Reference
Serrano ham slices Iberian ham slices	*Listeria innocua*	Total fluence (J/cm^2): 0, 2.1, 4.2 and 8.4 Peak power (J/cm^2/pulse): 0.3 Stored for 4 days at 20 °C	1 2	[43]
Pork loin	*Salmonella typhimurium* *Yersinia enterocolitica*	Total fluence (J/cm^2): 0.52–19.11 Peak power (J/cm^2/pulse): 1.27 Pulse width (μs): 300 Exposure time (s): 1–30 Distance from the lamp (cm): 8.3–13.4	0.4–1.71 0.39–1.69	[44]
Pork skin		Total fluence (J/cm^2): 19.11 Peak power (J/cm^2/pulse): 1.27 Pulse width (μs): 300 Exposure time (s): 30 Distance from the lamp (cm): 8.3	2.97 4.19	

A direct consequence of the reduction in spoilage microorganisms in foodborne products is an increase in the shelf-life of fresh products. The shelf-life of products may be extended by preventing deterioration and looking forward to maintaining organoleptic properties [45]. However, depending on the foodstuff and the PL treatment, a negative impact in organoleptic properties may appear; such is the case of the negative modification of sensory properties of bologna with 2.1 J/cm^2 treatment, while the use of 8.4 J/cm^2 did not produce sensory changes in cooked ham [46]. Products such as fresh-cut fruits and vegetables, fruit juices, meat, fish and derivative products (beef, tuna, salmon) are examples of extended shelf-life after the use of PL treatments [47]. The proper use of food preservation technologies and proper food packaging materials would contribute to extending the products shelf-life.

The use of PL is recommended for packaged foodstuff since this technology does not leave any residues, as is the case for the formation of hydrogen peroxide (H_2O_2) or peracetic acid (CH_3CO_3H) [24]. Nonetheless, there are contrary opinions in this regard; while some authors suggested the use of PL to decontaminate food packed products [47], there are other opinions that consider that there is a drawback in the use of this technology, since the packaging of foods treated with PL shall need aseptic conditions prior to packaging the product and then continue with the decontamination process [21]. This means that the whole process has to comply with sterilization standards for processing equipment and packaging containers to avoid cross-linked contamination of any kind.

4. Use of Pulsed Light for Grape Sanitation

The effectiveness of PL is higher when applied on surfaces than on liquids [48]. This entails that it is easier to apply the PL on grapes before crushing them to produce must or juice. The effect of PL as a treatment for the reduction in microbial populations in grapes has been evaluated, although the negative effect produced on microorganisms caused by the use of UV-C light, component also present in the PL flashes, was proven even before in various fruit matrices [49–51].

Despite the scarce experimental evidence on the use of PL on *Vitis vinifera* for sanitation purposes, there are data on laboratory-scale trials performed with two different energy doses, 300 and 600 J, providing an energy density of 1.1 and 2.1 J/cm^2, respectively [52]. The results obtained have shown effectiveness in the reduction in both yeast and bacteria naturally found on the surface of *Vitis vinifera* L. cv. Tempranillo. The treatment involved either 5 or 10 flashes (pulses) at each energy and the outcome has revealed more efficiency against bacteria populations, most probably due to being the largest population, when using the maximum energy possible regardless of the amount of pulses. Other studies have used different treatment set ups with less energy density on different fruit matrices. This is the case observed in trials intended for the elimination of *Botrytis cinerea* inoculated on strawberries [53]. One of the treatments involved pulsed light and combinations with heat and UV-C with an energy density of 0.05–0.1 J/cm^2 (0.5–1 kJ/m^2) in pulses of 40 and 120 s. PL at this low energy density did not affect the growth of mycelia, while combinations of PL and UV-C radiation delayed the spoilage caused by *B. cinerea* 24 h. It is then observed that flashes emitted during shorter periods of time and higher energy density would increase the effectiveness of the irradiation to avoid the development of spoilage organisms located on food surfaces.

The effect of PL is expected to be more effective on the outer layers of grapes where the pruina, a waxy film covering the berries, and the microorganisms are located. The skin of the berries is therefore expected to undergo any sort of damage as well. On this matter, Fava et al. have demonstrated that the UV-C light is capable of producing damages on the epicarp and the mesocarp of grapes [54]. The disruption caused by the treatment was observed on epidermal cell walls and even deeper, on collenquimatous subepidermal layers. As anthocyanins, red-like molecules responsible for the colour of grapes and other berries, are produced in the cytosol of epidermal cells of the berries and stored in vacuoles, the use of PL pulses is expected to increase the release of these colored molecules during maceration in winemaking. The transfer of pigments from the skin to the pulp has been documented for variety Tempranillo red grapes after PL treatment [55]; nonetheless, the visual effect observed in the berries does not have an influence on the pigment and phenolic content from wines without treatment.

Static treatments produced in batch-size laboratory cabinets, where grapes do not have free movement during flashes, reduce the frequency of damages on the epidermis and vacuoles of grapes. As a result, even though it seems that there is an increase in pigment extraction in grape musts after PL treatment, the analytical evaluation does not show statistical differences among treated and non-treated grapes [52]. The use of roller bed conveyor belts for PL sanitation of grapes may increase the incidence of disruptions on the epidermal cells of the berries with the potential increment of pigment extraction.

5. Influence of Pulsed Light on the Implantation of Non-*Saccharomyces* Yeasts in Musts

The elimination of native microbiota in fruits have industrial potential applications other than providing safe food for human consumption by the elimination of pathogens. In this way, biotechnological processes may be used not only to control the transformation of raw products into ready-to-eat meals, but also to design the organoleptic profile these products are to have. The sanitation of vinification grapes with PL enables the use of selected starters from commercial yeast and bacteria strains to ensure the population load and to reduce the use of SO_2 in wine [52,56]. Non-*Saccharomyces* yeasts are known for having low fermentation efficiency and having low ethanol yield [57–59]. They account for most of the yeast species found naturally on the skin of grapes (40 to 100), but their metabolic characteristics make them prone to disappearing at early fermentative phases in spontaneous or uninoculated fermentations, or when *Saccharomyces cerevisiae* strains are inoculated to assure their rapid dominance, and therefore the contribution of non-*Saccharomyces* yeasts to wine is minimized [60].

The so called non-conventional yeasts or non-*Saccharomyces* yeasts contribute to enhancing the aroma complexity in wines and are able to increase the yield in which desired fermentative compounds are produced [61,62]. The reasons attributed to non-*Saccharomyces* yeasts to perform this way are an increased production and releasing of enzymes and the high production of aromatic volatile compounds [63,64]. The enzymes involved in this enhancement include lipases, proteases, esterases and β-glycosidases; the aroma compounds are mainly esters and higher alcohols.

The reduction in native biota from the grape's pruina during grape sanitation with PL would allow winemakers to direct fermentations towards more customized wines. The reduction in microbiota is larger when PL is applied in comparison to sulphitation levels between 40 and 50 mg/L (Figure 4.). In a sulphited must, the reduction in wild yeasts would allow *S. cerevisiae* to implant and be the dominant species, since it is more resistant to SO_2 [65]. The fact that the reduction in yeasts is larger when applying PL favors the implantation of non-*Saccharomyces* yeasts.

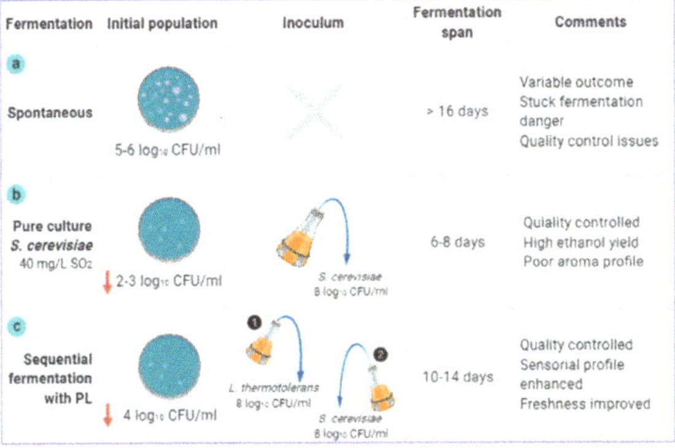

Figure 4. Summarized conditions comparison for three different fermentation scenarios: (**a**) spontaneous fermentation, (**b**) conventional pure culture fermentation with sulphited must and (**c**) sequential fermentation using grapes previously sanitized with PL treatment. Adapted from [52,60,62,66–69].

6. Repercussion of Pulsed Light in the Wine Freshness

In the context of global warming, there is a growing interest in improving the sensory perception of wines to counteract the harsh conditions of harvesting grapes in hot climate areas. Temperature changes and shifts in rainfall patterns would force the cultivars to thrive under such conditions. Higher temperatures will modify the chemistry of the grapes towards more sugar and less acid content, especially malic acid [70]. As a consequence, overripe grapes may produce wines with high alcohol content. The negative effects that global warming and climate change have on wines are perceived as inappropriate anthocyanin concentration affecting colour, imbalanced ratio of organic acids leading to tartaric acid addition to assure mouthfeel and microbial stability, and the potential production of odd overcooked aromas [71].

Actions to alleviate the negative impact of climate change include the use of elicitors applied to the canopy of vines. Elicitors, molecules able to activate secondary biosynthetic pathways in plants for self-protection, are used to try to reduce the differences in technological and physiological maturity of grapes by promoting a more rapid accumulation of phenolic compounds [72]. Another approach that promotes the formation of bioactive compounds is the use of PL in post-harvest products as an attempt to improve the quality of immature fruits such as tomatoes [73]. In this case, and contrary to what is aimed with the use of elicitors, the effect of PL is produced after the vegetative growth and once the fruits are harvested. Other approaches to diminish the negative impact of climate change involve the use of biotechnological solutions to help in reducing the alcohol content in wines, and therefore, the vinous or alcoholic perception of wines. These approaches consider using yeast strains with reduced glycolytic metabolism, yielding lower ethanol concentration in wines [59].

The freshness in wines is often perceived as a combination of parameters that all together contribute to increasing the fruity and floral scents and the acid character of wines. The aroma profile associated to freshness is produced by either fermentative metabolites, mainly esters, or by terpenic or thyolic precursors related to certain grape varieties released by yeast enzymatic activity [68]. Examples of such yeast strains cover the genera *Torulaspora*, *Wickerhamomyces*, *Metschnikowia*, *Lachancea* and *Hanseniaspora*, among others. Since these non-*Saccharomyces* yeast species have low–medium fermentative performance, the approach would consider mixed or sequential fermentations. Co-fermentations of *Saccharomyces cerevisiae* with *Hanseniaspora vineae* and *Metschnikowia pulcherrima* increases the total amount of acetate esters, ethyl esters and isoprenoids in wines [74]. The same *Hanseniaspora vineae* species has been proven to produce benzenoid compounds [75], as well as phenylpropanoid-derived compounds [76], capable of improving varietal and fermentative aromas of wines due to genetic variations in the enzymatic conformation in yeast strains. The fruity and floral contribution of these volatile compounds is noticed during the vinification process and so is the vanilla, woody or sweet coconut scents observed over the ageing period coming from hydroxybenzyl compounds produced from the metabolism of phenylpropanoids [75].

The acidic profile of wines can also be modified through yeast action, especially with the use of non-*Saccharomyces* yeast able to produce larger amounts of organic acids than conventional *Saccharomyces cerevisiae* strains commercially available. Such is the case of the yeast species *Lachancea thermotolerans*, known for its ability to consume and ferment glucose and fructose, and to assimilate galactose [77], but also known for its capability to produce lactic acid in a broad range of concentrations from 1 to 16.8 g/L [78]. This yeast species is able to reduce pH values in more than 0.5 units in a metabolic natural way. A reduction in pH values would also increase the effectiveness of molecular SO_2 added in lower dosages of total sulphites towards microbial stability in hot climate areas. Wines, mostly red wines and a few white wines, undergo malolactic fermentation (MLF) to reduce the amount of malic acid by a microbiological process performed by lactic acid bacteria strains of the genera *Lactobacillus* and *Oenococcus* [79,80]. The MLF usually takes place after the alcoholic fermentation (AF) and it usually needs special conditions for it to develop spontaneously; in most cases, the use of starter cultures is recommended. Among the reasons why the MLF is delayed or inhibited might be the concentration of lactic acid produced by yeast strains. High amounts of lactic acid produced by

L. thermotolerans may inhibit the growth of lactic acid bacteria such as *O. oeni* and therefore complicate the reduction in malic acid through MLF [81]. To counteract this drawback, the co-fermentation of *L. thermotolerans* and *O. oeni* has shown synergetic interactions towards the production of lactic acid through both metabolic pathways and, thus, achieving the reduction in pH values and the decrease in malic acid concentration in wines [82].

In terms of colour, the sensory profile can be influenced by the intensity and the colour hue. Red wines with red-brown hues are associated with oxidized processes and thus with aged or old wines [68]. On the contrary, bright red and blue-violet tones resemble young and fresher wines. In this way, colour may also contribute to creating a perception of freshness in wines. The production of colored molecules through the interaction of anthocyanins and metabolites during fermentation is also linked to yeast activity. Non-*Saccharomyces* yeasts contribute to the formation of pyranoanthocyanin pigments and oligomeric and polymeric adducts [64]. The formation of oligomers through the condensation of ethyl-bridged adducts of anthocyanins and flavan-3-ols produces molecules with absorption wavelengths of ca. 540 nm, towards red-blue hues [83]. The fermentation carried out completely with non-*Saccharomyces* yeasts seems to promote the formation of more oligomeric pigments than co-fermentations with *S. cerevisiae* [84]. There are several other molecular structures formed during ageing with wavelength absorbances higher than 530 nm, as in the case of A-type portisins with blue hues [85]. These last pigments involve acetaldehyde moieties in their molecular structure, vinyl linkages, favored by microbial activity through an increased production of this fermentative metabolite.

Taking into account the abovementioned contributions that non-*Saccharomyces* yeasts may have to modulate freshness, it is of great importance to assure and control the implantation of yeast strains, and eventually bacteria strains, capable of modifying the sensory profile of wines. An overview with the benefits of the use of PL at winemaking facilities is available in Figure 5. The possibility of applying PL in situ at a winemaking facility, with the use of automated roller bed conveyors, can ease the sanitation of grapes and the implementation of the hereinabove mentioned biotechnologies.

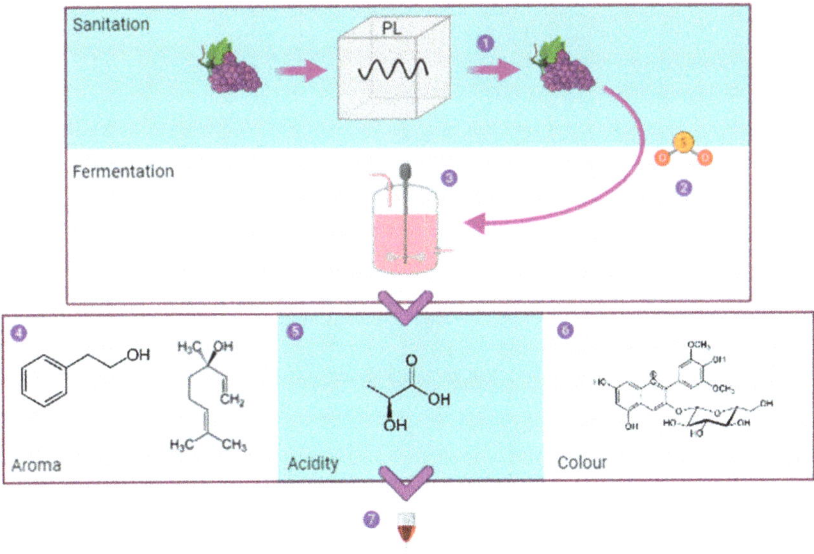

Figure 5. Summary of the potential benefits of using PL sanitation on grapes in the winemaking industry. **(1)** Inactivation or elimination of microbial population, **(2)** less usage of SO_2, **(3)** use of selected non-*Saccharomyces* yeasts, **(4)** enhance aroma profile, **(5)** modulate organic acidity, **(6)** increase pigment extraction and **(7)** gain freshness in wines towards customer preferences.

7. Conclusions

Pulsed light sanitation of grapes intended for use in winemaking production at the industrial scale may become affordable in the coming years. The efficiency of the UV-C light fraction of PL in inactivating microorganisms, the possibility of reducing the use of SO_2 to process and store musts, and the relatively low energy cost would make its deployment possible. Among the benefits observed in laboratory trials is the use of selected non-*Saccharomyces* strains from commercial traders to produce wines with enhanced organoleptic profiles in accordance with customers' demands and fulfilling quality control requirements.

One further challenge in the use of PL as non-thermal sanitation technology in the winemaking industry will be the design of systems able to process grape musts. The difficulties in achieving this will be to deal with a product with variable viscosity and with inhomogeneous particle size distribution. The energy needed for non-transparent liquids with low transmittance capacity and the scale-up design are issues that still have to be addressed. In this matter, the use of PL treatment to avoid the establishment of spoilage yeast during wine-making has to be evaluated with the use of special 1 mm width quartz cells. Additionally, further studies have to be performed in order to determine the impact that different PL doses have on the antioxidant capacity and the phenolic content of wines, as well as the effect of the treatment on other phenolic volatile compounds responsible for the wine aroma.

Author Contributions: A.S. drafted the manuscript; C.E. drafted the manuscript; I.L. supported the drafting of the manuscript and revise the manuscript; J.M.d.F. revised and corrected the manuscript; C.G. corrected the language and revised the manuscript; and A.M. revised the manuscript. All authors have read and agreed to the published version of the manuscript.

Funding: This research was funded by Ministerio de Ciencia, Innovación y Universidades grant number [RTI2018-096626-B-I00].

Conflicts of Interest: The authors declare no conflict of interest.

References

1. Barbosa-Canovas, G.V.; Schaffner, D.W.; Pierson, M.D.; Zhang, Q.H. Pulsed Light Technology. *J. Food Sci.* **2000**, *65*, 82–85. [CrossRef]
2. Johan, L.; Nilsson, A.; Gb, S.; Lefort, L.; Fr, L.; Hugh, J.; Price, V.; Gb, C. Pulsed light sources. U.S. Patent US6917631B2, 12 July 2005.
3. Unni, L.E.; Chauhan, O.P. Use of Pulsed Light in Food Processing. In *Non-Thermal Processing of Foods;* Chauhan, O.P., Ed.; CRC Press: Boca Raton, FL, USA, 2019; pp. 173–188.
4. Kirkpatrick, A.R. Method involving pulsed light processing of semiconductor devices. U.S. Patent US4151008, 24 April 1979.
5. Ryu, J.; Kim, H.S.; Hahn, H.T. Reactive sintering of copper nanoparticles using intense pulsed light for printed electronics. *J. Electron. Mater.* **2011**, *40*, 42–50. [CrossRef]
6. Babilas, P.; Schreml, S.; Szeimies, R.M.; Landthaler, M. Intense pulsed light (IPL): A review. *Lasers Surg. Med.* **2010**, *42*, 93–104. [CrossRef] [PubMed]
7. Negishi, K.; Tezuka, Y.; Kushikata, N.; Wakamatsu, S. Photorejuvenation for Asian skin by intense pulsed light. *Dermatol. Surg.* **2001**, *27*, 627–632. [PubMed]
8. Goldberg, D.J. New collagen formation after dermal remodeling with an intense pulsed light source. *J. Cutan. Laser Ther.* **2000**, *2*, 59–61. [CrossRef] [PubMed]
9. Schroeter, C.A.; Below, S.H.-V.; Neumann, H.A.M. Effective treatment of rosacea using intense pulsed light systems. *Dermatol. Surg.* **2005**, *31*, 1285–1289. [CrossRef]
10. Angermeier, M.C. Treatment of facial vascular lesions with intense pulsed light. *J. Cosmet. Laser Ther.* **1999**, *1*, 95–100. [CrossRef]
11. Hernández-Pérez, E.; Colombo-Charrier, E.; Valencia-Ibiett, E. Intense pulsed light in the treatment of striae distensae. *Dermatol. Surg.* **2002**, *28*, 1124–1130.
12. Gold, M.H.; Bell, M.W.; Foster, T.D.; Street, S. Long-term epilation using the EpiLight broad band, intense pulsed light Hair Removal System. *Dermatol. Surg.* **1997**, *23*, 909–913. [CrossRef]

13. Pollock, A.M.; Singh, A.P.; Ramaswamy, H.S.; Ngadi, M.; Singh, P. Pulsed light destruction kinetics of *L. monocytogenes*. *LWT-Food Sci. Technol.* **2017**, *84*, 114–121. [CrossRef]
14. MacGregor, S.J.; Rowan, N.J.; McIlvaney, L.; Anderson, J.G.; Fouracre, R.A.; Farish, O. Light inactivation of food-related pathogenic bacteria using a pulsed power source. *Lett. Appl. Microbiol.* **1998**, *27*, 67–70. [CrossRef] [PubMed]
15. Rowan, N.J.; MacGregor, S.J.; Anderson, J.G.; Fouracre, R.A.; McIlvaney, L.; Farish, O. Pulsed-Light Inactivation of Food-Related Microorganisms. *Appl. Environ. Microbiol.* **1999**, *65*, 1312–1315. [CrossRef] [PubMed]
16. Elmnasser, N.; Guillou, S.; Leroi, F.; Orange, N.; Bakhrouf, A.; Federighi, M. Pulsed-light system as a novel food decontamination technology: A review. *Can. J. Microbiol.* **2007**, *53*, 813–821. [CrossRef]
17. Brown, A.C. *Understanding Food-Principles and Preparation*, 3rd ed.; Thomson Wadsworth: Belmont, CA, USA, 2008; ISBN 9780495107453.
18. Song, K.; Taghipour, F.; Mohseni, M. Microorganisms inactivation by wavelength combinations of ultraviolet light-emitting diodes (UV-LEDs). *Sci. Total Environ.* **2019**, *665*, 1103–1110. [CrossRef] [PubMed]
19. Gómez-López, V.M.; Ragaert, P.; Debevere, J.; Devlieghere, F. Pulsed light for food decontamination: A review. *Trends Food Sci. Technol.* **2007**, *18*, 464–473. [CrossRef]
20. Wiktor, A.; Mandal, R.; Singh, A.; Singh, A.P. Pulsed Light treatment below a Critical Fluence. *Foods* **2019**, *8*, 1–13.
21. Martín-belloso, O.; Soliva-fortuny, R.; Elez-Martínez, P.; Marsellés-Fontanet, R.; Vega-mercado, H. Non-thermal Processing Technologies. In *Food Safety Management*; Motarjemi, Y., Lelieveld, H., Eds.; Academic Press Inc.: London, UK, 2014; pp. 443–465. ISBN 9780123815040.
22. Rowan, N.J. Pulsed light as an emerging technology to cause disruption for food and adjacent industries–Quo vadis? *Trends Food Sci. Technol.* **2019**, *88*, 316–332. [CrossRef]
23. Wekhof, A. Disinfection with flash lamps. *PDA J. Pharm. Sci. Technol.* **2000**, *54*, 264–276.
24. Kumar, P.; Han, J.H. Packaging materials for non-thermal processing of food and beverages. In *Emerging Food Packaging Technologies*; Yan, K., Lee, D.S., Eds.; Woodhead Publishing: Oxford, UK, 2012; pp. 323–334.
25. Morris, C.; Brody, A.L.; Wicker, L. Non-Thermal Food Processing/Preservation Technologies: A Review with Packaging Implications. *Packag. Tech. Sci.* **2007**, *20*, 275–286. [CrossRef]
26. Agüero, M.V.; Jagus, R.J.; Martín-Belloso, O.; Soliva-Fortuny, R. Surface decontamination of spinach by intense pulsed light treatments: Impact on quality attributes. *Postharvest Biol. Technol.* **2016**, *121*, 118–125. [CrossRef]
27. Llano, K.R.A.; Marsellés-Fontanet, A.R.; Martín-Belloso, O.; Soliva-Fortuny, R. Impact of pulsed light treatments on antioxidant characteristics and quality attributes of fresh-cut apples. *Innov. Food Sci. Emerg. Technol.* **2016**, *33*, 206–215. [CrossRef]
28. Valdivia-Nájar, C.G.; Martín-Belloso, O.; Giner-Seguí, J.; Soliva-Fortuny, R. Modeling the Inactivation of *Listeria innocua* and *Escherichia coli* in Fresh-Cut Tomato Treated with Pulsed Light. *Food Bioprocess. Technol.* **2017**, *10*, 266–274. [CrossRef]
29. Leng, J.; Mukhopadhyay, S.; Sokorai, K.; Ukuku, D.O.; Fan, X.; Olanya, M.; Juneja, V. Inactivation of *Salmonella* in cherry tomato stem scars and quality preservation by pulsed light treatment and antimicrobial wash. *Food Control.* **2020**, *110*, 107005. [CrossRef]
30. Ignat, A.; Manzocco, L.; Maifreni, M.; Bartolomeoli, I.; Nicoli, M.C. Surface decontamination of fresh-cut apple by pulsed light: Effects on structure, colour and sensory properties. *Postharvest Biol. Technol.* **2014**, *91*, 122–127. [CrossRef]
31. Tao, T.; Ding, C.; Han, N.; Cui, Y.; Liu, X.; Zhang, C. Evaluation of pulsed light for inactivation of foodborne pathogens on fresh-cut lettuce: Effects on quality attributes during storage. *Food Packag. Shelf Life* **2019**, *21*, 100358. [CrossRef]
32. Keklik, N.M.; Demirci, A.; Puri, V.M. Decontamination of unpackaged and vacuum-packaged boneless chicken breast with pulsed ultraviolet light. *Poult. Sci.* **2010**, *89*, 570–581. [CrossRef]
33. Xu, W.; Wu, C. The impact of pulsed light on decontamination, quality, and bacterial attachment of fresh raspberries. *Food Microbiol.* **2016**, *57*, 135–143. [CrossRef]
34. Aguiló-Aguayo, I.; Charles, F.; Renard, C.M.G.C.; Page, D.; Carlin, F. Pulsed light effects on surface decontamination, physical qualities and nutritional composition of tomato fruit. *Postharvest Biol. Technol.* **2013**, *86*, 29–36. [CrossRef]

35. Aguiló-Aguayo, I.; Oms-Oliu, G.; Martín-Belloso, O.; Soliva-Fortuny, R. Impact of pulsed light treatments on quality characteristics and oxidative stability of fresh-cut avocado. *LWT-Food Sci. Technol.* **2014**, *59*, 320–326. [CrossRef]
36. Valdivia-Nájar, C.G.; Martín-Belloso, O.; Soliva-Fortuny, R. Impact of pulsed light treatments and storage time on the texture quality of fresh-cut tomatoes. *Innov. Food Sci. Emerg. Technol.* **2018**, *45*, 29–35. [CrossRef]
37. Taştan, Ö.; Pataro, G.; Donsì, F.; Ferrari, G.; Baysal, T. Decontamination of fresh-cut cucumber slices by a combination of a modified chitosan coating containing carvacrol nanoemulsions and pulsed light. *Int. J. Food Microbiol.* **2017**, *260*, 75–80. [CrossRef] [PubMed]
38. Ramos-Villarroel, A.Y.; Martín-Belloso, O.; Soliva-Fortuny, R. Combined effects of malic acid dip and pulsed light treatments on the inactivation of *Listeria innocua* and *Escherichia coli* on fresh-cut produce. *Food Control.* **2015**, *52*, 112–118. [CrossRef]
39. Huang, Y.; Ye, M.; Cao, X.; Chen, H. Pulsed light inactivation of murine norovirus, Tulane virus, *Escherichia coli* O157:H7 and *Salmonella* in suspension and on berry surfaces. *Food Microbiol.* **2017**, *61*, 1–4. [CrossRef] [PubMed]
40. Cao, X.; Huang, R.; Chen, H. Evaluation of pulsed light treatments on inactivation of *Salmonella* on blueberries and its impact on shelf-life and quality attributes. *Int. J. Food Microbiol.* **2017**, *260*, 17–26. [CrossRef] [PubMed]
41. Xu, W.; Chen, H.; Wu, C. *Salmonella* and *Escherichia coli* O157:H7 inactivation, color, and bioactive compounds enhancement on raspberries during frozen storage after decontamination using new formula sanitizer washing or pulsed light. *J. Food Prot.* **2016**, *79*, 1107–1114. [CrossRef] [PubMed]
42. Xu, W.; Chen, H.; Huang, Y.; Wu, C. Decontamination of *Escherichia coli* O157: H7 on green onions using pulsed light (PL) and PL-surfactant-sanitizer combinations. *Int. J. Food Microbiol.* **2013**, *166*, 102–108. [CrossRef]
43. Fernández, M.; Hospital, X.F.; Cabellos, C.; Hierro, E. Effect of pulsed light treatment on *Listeria* inactivation, sensory quality and oxidation in two varieties of Spanish dry-cured ham. *Food Chem.* **2020**, *316*, 126294. [CrossRef]
44. Koch, F.; Wiacek, C.; Braun, P.G. Pulsed light treatment for the reduction of *Salmonella Typhimurium* and *Yersinia enterocolitica* on pork skin and pork loin. *Int. J. Food Microbiol.* **2019**, *292*, 64–71. [CrossRef]
45. Vaclavik, V.A.; Christian, E.W. *Essentials of Food Science*, 4th ed.; Heldman, D.R., Ed.; Springer: New York, NY, USA, 2008; Volume 45, ISBN 9781461491378.
46. Marriott, N.G.; Gravani, R.B. *Principles of Food Sanitation*, 5th ed.; Springer: New York, NY, USA, 2006; ISBN 9780387250250.
47. Mahendran, R.; Ramanan, K.R.; Barba, F.J.; Lorenzo, J.M.; López-Fernández, O.; Munekata, P.E.S.; Roohinejad, S.; Sant'Ana, A.S.; Tiwari, B.K. Recent advances in the application of pulsed light processing for improving food safety and increasing shelf life. *Trends Food Sci. Technol.* **2019**, *88*, 67–79. [CrossRef]
48. Oms-Oliu, G.; Martín-Belloso, O.; Soliva-Fortuny, R. Pulsed light treatments for food preservation. A review. *Food Bioprocess. Technol.* **2010**, *3*, 13–23. [CrossRef]
49. Guerrero-Beltrán, J.Á.; Welti-Chanes, J.; Barbosa-Cánovas, G.V. Ultraviolet-C light processing of grape, cranberry and grapefruit juices to inactivate *Saccharomyces cerevisiae*. *J. Food Process. Eng.* **2009**, *32*, 916–932. [CrossRef]
50. Valero, A.; Begum, M.; Leong, S.L.; Hocking, A.D.; Ramos, A.J.; Sanchis, V.; Marín, S. Effect of germicidal UVC light on fungi isolated from grapes and raisins. *Lett. Appl. Microbiol.* **2007**, *45*, 238–243. [CrossRef] [PubMed]
51. Pala, Ç.U.; Toklucu, A.K. Effects of UV-C Light Processing on Some Quality Characteristics of Grape Juices. *Food Bioprocess. Technol.* **2013**, *6*, 719–725. [CrossRef]
52. Escott, C.; Vaquero, C.; del Fresno, J.M.; Bañuelos, M.A.; Loira, I.; Han, S.; Bi, Y.; Morata, A.; Suárez-Lepe, J.A. Pulsed light effect in red grape quality and fermentation. *Food Bioprocess. Technol.* **2017**, *10*, 1540–1547. [CrossRef]
53. Marquenie, D.; Michiels, C.W.; Van Impe, J.F.; Schrevens, E.; Nicolaï, B.N. Pulsed white light in combination with UV-C and heat to reduce storage rot of strawberry. *Postharvest Biol. Technol.* **2003**, *28*, 455–461. [CrossRef]
54. Fava, J.; Hodara, K.; Nieto, A.; Guerrero, S.; Alzamora, S.M.; Castro, M.A. Structure (micro, ultra, nano), color and mechanical properties of Vitis labrusca L. (grape berry) fruits treated by hydrogen peroxide, UV-C irradiation and ultrasound. *Food Res. Int.* **2011**, *44*, 2938–2948. [CrossRef]

55. Morata, A.; Loira, I.; Vejarano, R.; González, C.; Callejo, M.J.; Suárez-Lepe, J.A. Emerging preservation technologies in grapes for winemaking. *Trends Food Sci. Technol.* **2017**, *67*, 36–43. [CrossRef]
56. Morata, A.; Loira, I.; Guamis, B.; Raso, J.; del Fresno, J.M.; Escott, C.; Bañuelos, M.A.; Álvarez, I.; Tesfaye, W.; González, C.; et al. Emerging technologies to increase extraction, control microorganisms, and reduce SO2. In *Winemaking-Stabilization, Aging Chemistry and Biochemistry*; Cosme, F., Ed.; IntechOpen: London, UK, 2020; pp. 1–20.
57. Canonico, L.; Comitini, F.; Oro, L.; Ciani, M. Sequential Fermentation with Selected Immobilized Non-*Saccharomyces* Yeast for Reduction of Ethanol Content in Wine. *Front. Microbiol.* **2016**, *7*, 278–289. [CrossRef]
58. Contreras, A.; Hidalgo, C.; Schmidt, S.; Henschke, P.A.; Curtin, C.; Varela, C. The application of non-*Saccharomyces* yeast in fermentations with limited aeration as a strategy for the production of wine with reduced alcohol content. *Int. J. Food Microbiol.* **2015**, *205*, 7–15. [CrossRef]
59. Loira, I.; Morata, A.; González, C.; Suárez-Lepe, J.A. Selection of Glycolytically Inefficient Yeasts for Reducing the Alcohol Content of Wines from Hot Regions. *Food Bioprocess. Technol.* **2012**, *5*, 2787–2796. [CrossRef]
60. Bisson, L.F.; Joseph, C.M.L.; Domizio, P. Yeasts. In *Biology of Microorganisms on Grapes, in Must and in Wine*; König, H., Unden, G., Fröhlich, J., Eds.; Springer International Publishing AG: Cham, Switzerland, 2017; pp. 65–102. ISBN 9783319600208.
61. Rêgo, E.S.B.; Rosa, C.A.; Freire, A.L.; Machado, A.M.d.R.; Gomes, F.d.C.O.; Costa, A.S.P.d.; Mendonça, M.d.C.; Hernández-Macedo, M.L.; Padilha, F.F. Cashew wine and volatile compounds produced during fermentation by non-*Saccharomyces* and Saccharomyces yeast. *LWT* **2020**, *126*, 109291. [CrossRef]
62. Del Fresno, J.M.; Morata, A.; Loira, I.; Bañuelos, M.A.; Escott, C.; Benito, S.; Chamorro, C.G.; Suárez-Lepe, J.A. Use of non-*Saccharomyces* in single-culture, mixed and sequential fermentation to improve red wine quality. *Eur. Food Res. Technol.* **2017**, *243*, 2175–2185. [CrossRef]
63. Liu, P.T.; Lu, L.; Duan, C.Q.; Yan, G.L. The contribution of indigenous non-*Saccharomyces* wine yeast to improved aromatic quality of Cabernet Sauvignon wines by spontaneous fermentation. *LWT-Food Sci. Technol.* **2016**, *71*, 356–363. [CrossRef]
64. Morata, A.; Escott, C.; Loira, I.; del Fresno, J.M.; González, C.; Suárez-Lepe, J.A. Influence of *Saccharomyces* and non-*Saccharomyces* Yeasts in the Formation of Pyranoanthocyanins and Polymeric Pigments during Red Wine Making. *Molecules* **2019**, *24*, 4490. [CrossRef] [PubMed]
65. Suárez-Lepe, J.A.; Morata, A. New trends in yeast selection for winemaking. *Trends Food Sci. Technol.* **2012**, *23*, 39–50. [CrossRef]
66. Pateraki, C.; Paramithiotis, S.; Doulgeraki, A.I.; Kallithraka, S.; Kotseridis, Y.; Drosinos, E.H. Effect of sulfur dioxide addition in wild yeast population dynamics and polyphenolic composition during spontaneous red wine fermentation from Vitis vinifera cultivar Agiorgitiko. *Eur. Food Res. Technol.* **2014**, *239*, 1067–1075. [CrossRef]
67. Martins, V.; Lopez, R.; Garcia, A.; Teixeira, A.; Gerós, H. Vineyard calcium sprays shift the volatile profile of young red wine produced by induced and spontaneous fermentation. *Food Res. Int.* **2020**, *131*, 108983. [CrossRef] [PubMed]
68. Morata, A.; Escott, C.; Bañuelos, M.A.; Loira, I.; Del Fresno, J.M.; González, C.; Suárez-lepe, J.A. Contribution of non-*Saccharomyces* yeasts to wine freshness. A review. *Biomolecules* **2020**, *10*, 34. [CrossRef]
69. Delsart, C.; Grimi, N.; Boussetta, N.; Sertier, C.M.; Ghidossi, R.; Peuchot, M.M.; Vorobiev, E. Comparison of the effect of pulsed electric field or high voltage electrical discharge for the control of sweet white must fermentation process with the conventional addition of sulfur dioxide. *Food Res. Int.* **2015**, *77*, 718–724. [CrossRef]
70. Mozell, M.R.; Thachn, L. The impact of climate change on the global wine industry: Challenges & solutions. *Wine Econ. Policy* **2014**, *3*, 81–89.
71. Keller, M. Managing grapevines to optimise fruit development in a challenging environment: A climate change primer for viticulturists. *Aust. J. Grape Wine Res.* **2010**, *16*, 56–69. [CrossRef]
72. Portu, J.; López, R.; Baroja, E.; Santamaría, P.; Garde-Cerdán, T. Improvement of grape and wine phenolic content by foliar application to grape-vine of three different elicitors: Methyl jasmonate, chitosan, and yeast extract. *Food Chem.* **2016**, *201*, 213–221. [CrossRef] [PubMed]

73. Pataro, G.; Sinik, M.; Capitoli, M.M.; Donsì, G.; Ferrari, G. The influence of post-harvest UV-C and pulsed light treatments on quality and antioxidant properties of tomato fruits during storage. *Innov. Food Sci. Emerg. Technol.* **2015**, *30*, 103–111. [CrossRef]
74. Zhang, B.-Q.; Shen, J.-Y.; Duan, C.-Q.; Yan, G.-L. Use of indigenous *Hanseniaspora vineae* and *Metschnikowia pulcherrima* co-fermentation with *Saccharomyces cerevisiae* to improve the aroma diversity of Vidal Blanc icewine. *Front. Microbiol.* **2018**, *9*, 2303. [CrossRef]
75. Martin, V.; Giorello, F.; Fariña, L.; Minteguiaga, M.; Salzman, V.; Boido, E.; Aguilar, P.S.; Gaggero, C.; Dellacassa, E.; Mas, A.; et al. De novo synthesis of benzenoid compounds by the yeast Hanseniaspora vineae increases the flavor diversity of wines. *J. Agric. Food Chem.* **2016**, *64*, 4574–4583. [CrossRef] [PubMed]
76. Giorello, F.; Valera, M.J.; Martin, V.; Parada, A.; Salzman, V.; Camesasca, L.; Fariña, L.; Boido, E.; Medina, K.; Dellacassa, E.; et al. Genomic and transcriptomic basis of *Hanseniaspora vineae*'s impact on flavor diversity and wine quality. *Appl. Environ. Microbiol.* **2019**, *85*, 1–43.
77. Senses-Ergul, S.; Ágoston, R.; Belák, Á.; Deák, T. Characterization of some yeasts isolated from foods by traditional and molecular tests. *Int. J. Food Microbiol.* **2006**, *108*, 120–124. [CrossRef] [PubMed]
78. Morata, A.; Loira, I.; Tesfaye, W.; Bañuelos, M.; González, C.; Lepe, J.S. *Lachancea thermotolerans* applications in wine technology. *Fermentation* **2018**, *4*, 53. [CrossRef]
79. Lucio, O.; Pardo, I.; Heras, J.M.; Krieger-Weber, S.; Ferrer, S. Use of starter cultures of *Lactobacillus* to induce malolactic fermentation in wine. *Aust. J. Grape Wine Res.* **2017**, *23*, 15–21. [CrossRef]
80. Krieger-Weber, S.; Heras, J.M.; Suarez, C. *Lactobacillus plantarum*, a New Biological Tool to Control Malolactic Fermentation: A Review and an Outlook. *Beverages* **2020**, *6*, 23. [CrossRef]
81. Morata, A.; Bañuelos, M.A.; López, C.; Song, C.; Vejarano, R.; Loira, I.; Palomero, F.; Lepe, J.A.S. Use of fumaric acid to control pH and inhibit malolactic fermentation in wines. *Food Addit. Contam.-Part A Chem. Anal. Control. Expo. Risk Assess.* **2020**, *37*, 228–238. [CrossRef] [PubMed]
82. Morata, A.; Bañuelos, M.A.; Vaquero, C.; Loira, I.; Cuerda, R.; Palomero, F.; González, C.; Suárez-Lepe, J.A.; Wang, J.; Han, S.; et al. *Lachancea thermotolerans* as a tool to improve pH in red wines from warm regions. *Eur. Food Res. Technol.* **2019**, *245*, 885–894. [CrossRef]
83. Escott, C.; Morata, A.; Loira, I.; Tesfaye, W.; Suarez-Lepe, J.A. Characterization of polymeric pigments and pyranoanthocyanins formed in microfermentations of non-*Saccharomyces* yeasts. *J. Appl. Microbiol.* **2016**, *121*, 1346–1356. [CrossRef] [PubMed]
84. Escott, C.; Del Fresno, J.M.; Loira, I.; Morata, A.; Tesfaye, W.; González, C.; Suárez-Lepe, J.A. Formation of polymeric pigments in red wines through sequential fermentation of flavanol-enriched musts with non-*Saccharomyces* yeasts. *Food Chem.* **2018**, *239*, 975–983. [CrossRef] [PubMed]
85. Oliveira, J.; de Freitas, V.; Mateus, N. Polymeric Pigments in Red Wines. In *Red Wine Technology*; Morata, A., Ed.; Academic Press: London, UK, 2019; pp. 207–218, ISBN 9780128144008.

Review

Lactobacillus plantarum, a New Biological Tool to Control Malolactic Fermentation: A Review and an Outlook

Sibylle Krieger-Weber [1,*], José María Heras [2] and Carlos Suarez [2]

1 Lallemand, Office Korntal-Münchingen, In den Seiten 53, 70825 Korntal-Münchingen, Germany
2 Lallemand Spain-Portugal, c/ Tomás Edison nº 4, 28521 Madrid, Spain; jmheras@lallemand.com (J.M.H.); csuarez@lallemand.com (C.S.)
* Correspondence: skrieger@lallemand.com; Tel.: +49-172-7120765

Received: 26 February 2020; Accepted: 1 April 2020; Published: 7 April 2020

Abstract: Malolactic fermentation (MLF) in wine is an important step in the vinification of most red and some white wines, as stands for the biological conversion of L-malic acid into L-lactic acid and carbon dioxide, resulting in a decrease in wine acidity. MLF not only results in a biological deacidification, it can exert a significant impact on the organoleptic qualities of wine. This paper reviews the biodiversity of lactic acid bacteria (LAB) in wine, their origin, and the limiting conditions encountered in wine, which allow only the most adapted species and strains to survive and induce malolactic fermentation. Of all the species of wine LAB, *Oenococcus oeni* is probably the best adapted to overcome the harsh environmental wine conditions and therefore represents the majority of commercial MLF starter cultures. Wine pH is most challenging, but, as a result of global warming, *Lactobacillus* sp. is more often reported to predominate and be responsible for spontaneous malolactic fermentation. Some *Lactobacillus plantarum* strains can tolerate the high alcohol and SO_2 levels normally encountered in wine. This paper shows the potential within this species for the application as a starter culture for induction of MLF in juice or wine. Due to its complex metabolism, a range of compositional changes can be induced, which may positively affect the quality of the final product. An example of a recent isolate has shown most interesting results, not only for its capacity to induce MLF after direct inoculation, but also for its positive contribution to the wine quality. Degrading hexose sugars by the homo-fermentative pathway, which poses no risk of acetic acid production from the sugars, is an interesting alternative to control MLF in high pH wines. Within this species, we can expect more strains with interesting enological properties.

Keywords: malolactic fermentation; *Lactobacillus plantarum*; *Oenococcus oeni*; facultative hetero-fermentative; starter cultures

---

1. Introduction

Malolactic fermentation (MLF), the process of biological de-acidification in winemaking, is based on the L-malic acid decarboxylation to L-lactic acid and CO_2. It can occur during or after alcoholic fermentation as a result of the metabolic activity of lactic acid bacteria (LAB), which are present in wine at all stages of winemaking. Four genera were identified as the principal organisms involved in MLF: *Lactobacillus*, *Leuconostoc*, *Oenococcus*, and *Pediococcus* [1]. Wine pH is most selective, and, at a pH below 3.5, generally only strains of *Oenococcus oeni* can survive and express malolactic activity. *O. oeni* is probably the best adapted to overcome the harsh environmental wine conditions and therefore most of the commercial MLF starter cultures consist of strains from this species. Traditionally, when selected wine bacteria are used, inoculation is performed at the completion of alcoholic fermentation (AF). Since 1980, researchers have explored the possibility of inoculating wine LAB into the grape must together with the yeast or shortly after the yeast at the beginning of the alcoholic fermentation. Today, we have identified two different timings throughout the winemaking process for inoculating wine LAB into the wine: co-inoculation with yeast (selected wine bacteria added 24 to 72 h after yeast addition) or sequential inoculation, when selected wine lactic acid bacteria are added at the end of, or just after the completion of, AF.

Wine pH has been increasing gradually for the last several years. Red wines with pHs over 3.5–3.6 are more and more frequent. At these pH levels, we can observe very fast growth of various indigenous microorganisms, some of which are spoilage bacteria that can cause loss of wine quality. Among these species, *Lactobacillus plantarum* strains have shown most interesting results for their capacity to induce MLF under high pH conditions, their facultative hetero-fermentative properties that avoid acetic acid production from hexose sugars and their more complex enzymatic profile and different metabolism compared to *O. oeni*, which could play an important role in the modification of wine aromas.

Besides pH, the ethanol produced by the yeast during alcoholic fermentation is another limiting factor for bacterial growth and survival in wine. Radler [2], Peynaud and Domercq [3], and Henick-Kling [4] reported an increasing inhibition above 5% (*v/v*). The degree of ethanol tolerance is, however, strain dependent. Specific details of alcohol sensitivity for the various species of wine LAB are contradicting. Davis et al. [5] reported strains of *Lactobacillus* and *Pediococcus* being in general more tolerant to high ethanol concentrations than *O. oeni*. In contrast to this, Henick-Kling [6] reported *O. oeni* being only partially inhibited by ethanol concentrations above 5% (*v/v*) and able to tolerate up to 14% (*v/v*) alcohol, while the growth of *L. plantarum* stops at ethanol concentrations of 5–6% (*v/v*). The first *L. plantarum* starter culture was introduced in the late 1980s in the United States and later also in Europe. Prahl et al. [7,8] proposed to inoculate the grape juice before alcoholic fermentation using a facultative hetero-fermentative *L. plantarum* starter culture. In EP0398957B1 [7], they disclosed a method of introducing an important freeze-dried biomass of *L. plantarum* directly into must or fruit juice to induce MLF without significant consumption of sugars present in the must or fruit juice and substantially without any production of volatile acidity. This malolactic bacteria strain had little alcohol tolerance and had been unable to survive in the fermented wine. The application of this culture had only been recommended for partial malic acid degradation in low pH white wines.

In 2004, Bou and Krieger [9] filed a patent on "Alcohol-tolerant malolactic strains for the maturation of wines with average or high pH" under the international application number PCT/FR2004/001421. The patent relates to alcohol tolerant LAB strains of the genera *Lactobacillus* and *Pediococcus* capable of initiating and carrying out a complete MLF upon direct inoculation in dried, frozen, or lyophilized state into a wine with an alcohol content of 10% (*v/v*) or more and an average high pH level. Dating back to 2005, a new selection of *L. plantarum* at Università Cattolica del Sacro Cuore in Italy resulted in a very effective *L. plantarum* culture V22, adapted to high pH wines, and showing good alcohol tolerance [10].

In 2012, Soerensen et al. [11] filed a patent application on "*Lactobacillus plantarum* cells with improved resistance to high concentrations of ethanol" (WO 2012/17200). The invention relates to cycloserine resistant mutants of lactic acid bacteria having improved resistance towards ethanol. The cycloserine resistant mutants of lactic acid bacteria had been proposed for use to induce malolactic fermentation in wine having high alcohol levels, but, as outlined above, alcohol tolerant *L. plantarum* strains can also be isolated from nature.

In 2016, a new highly concentrated *Lactobacillus plantarum* starter culture was introduced to the markets [12]. The new starter culture, called ML Prime™, is issued from an optimized process that promotes very high malolactic activity as soon as it is added to must. Despite the good alcohol tolerance of this pure *Lactobacillus plantarum* culture with the homo-fermentative metabolism of hexose sugars, its most interesting application is in co-inoculation (inoculation 24 h after the wine yeast) without any risk of volatile acidity production during MLF even under high pH conditions. Due to the very high malolactic enzymatic activity and early inoculation shortly after the selected wine yeast into the fermenting must, MLF is therefore completed in record time (3–7 days) during alcoholic fermentation.

This way, wines can be stabilized early and protected from further contamination and thus retain their sensory integrity. More recently, this *L. plantarum* starter culture ML Prime™ had been also proposed for a specific application in white wine to achieve a partial malolactic fermentation under lower pH conditions. In the white wine application, co- and sequential inoculation can be applied.

The majority of the selected wine lactic acid bacteria starter cultures are the pure single strain cultures, but, in 2008, the Institute for Wine Biotechnology at Stellenbosch University [13] launched a project to study on the possible application of mixed MLF starter cultures consisting of one selected *L. plantarum* and one selected *O. oeni* strain deriving from the Stellenbosch strain collection. In 2010, the first mixed LAB species culture was as Co-Inoculant NT202 and proposed for simultaneous inoculation together with a specific yeast strain NT202. The authors claim the importance of the *L. plantarum* strain in the mix for its sensory contribution and of the *O. oeni* strain for its malolactic enzyme activity. Certain strains within the *L. plantarum* species have been found to possess even more diverse enzymatic activities, which could contribute to the wine aroma profile than *O. oeni*.

2. Biodiversity of Lactic Acid Bacteria in Wine

Winemaking is a microbiological process involving a very complex system and it involves numerous microbial transformations comprising a complex succession of various yeast and bacterial species. Malolactic fermentation (MLF) can occur during or after alcoholic fermentation and is carried out by one or more species of lactic acid bacteria (Figure 1).

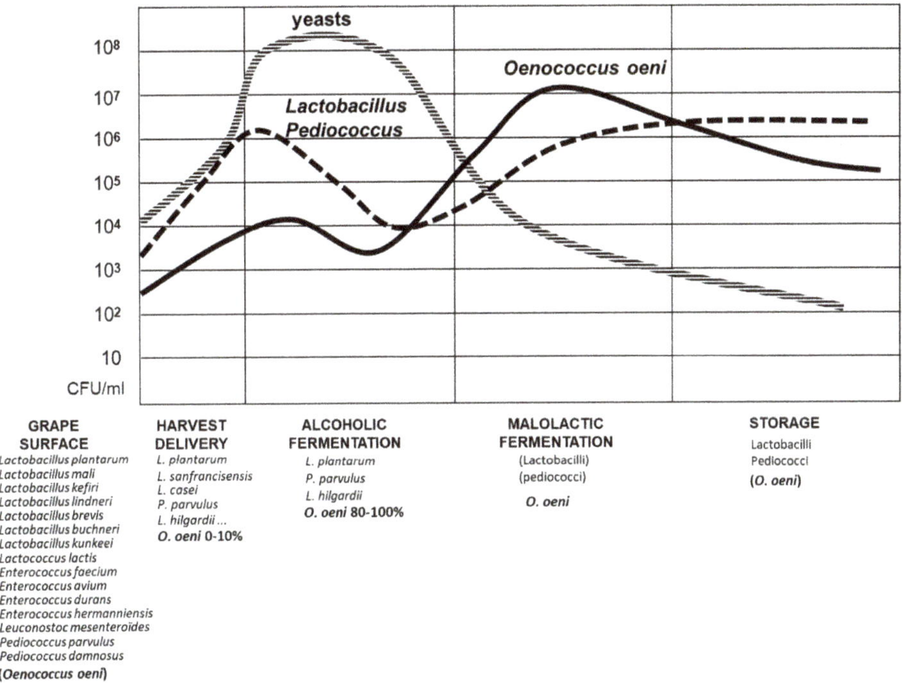

Figure 1. Population kinetics of wine lactic acid bacteria from the vineyard to the wine (Modified from Patrick Lucas, 2016, International ML School Lallemand Toulouse).

Different LAB enter into grape juice/wine from the surface of grape berries, stems, leaves, and soil and winery equipment. In the vineyard, LAB species diversity associated with grape surfaces is rather limited, mainly due to their nutritional requirements [14,15]. The population density of LAB is very limited, especially in comparison to the indigenous yeast population found on grapes [16].

Pediococcus, *Leuconostoc*, and *Lactobacillus* species occur on grapes more frequently than *O. oeni* [17]. In addition to grape surfaces, bacterial strains can also be isolated from the cellar environment, such as fermentation tanks and barrels and poorly sanitized winery equipment, such as pipes and valves [17,18]. Shortly after crushing and the start of AF, the LAB population in the grape must generally range from 10^3 to 10^4 cfu/mL (colony forming units per milliliter), and the LAB species largely belonging to the species of *Lactobacillus* and *Pediococcus* disappear progressively during the AF [19]. The decrease could be attributed to increased ethanol concentrations, high SO_2 concentrations, initial low pH, low temperatures, the nutrient depletion, and/or competitive interactions with the yeast culture [16,20]. During spontaneous MLF, *O. oeni* is the major bacterial species found, however, several species can be occasionally detected, mainly *Lactobacillus*, *Pediococcus*, and *Leuconostoc* [1,19]. In some of the warmer wine growing regions, *L. plantarum* is more frequently isolated from spontaneous malolactic fermentations [13,21–23]. Lerm et al. reported three *O. oeni* and three *L. plantarum strains* from South Africa wine isolates for use as MLF starter cultures. Bergeral et al. [21] had studied the properties of *Lactobacillus plantarum* strains isolated from grape must fermentation Apulian wines in order to select suitable starter for MLF, and Valdés La Hens et al. [22] reported the Prevalence of *L. plantarum* and *O. oeni* during spontaneous fermentation in Patagonian red wines. More recently, Lópes-Seijas et al. [23] evaluated malolactic bacteria associated with wine from the Albariño variety. Different to what has been described from other wine growing regions, the predominant species in the region of Val do Salnés in Spain were *L. hilgardii*, *L.* paracasei, and *L.* plantarum. Nevertheless *O. oeni* is most frequently the predominant species at the later stages of vinification (Figure 1), since it is best adapted to the limiting conditions encountered in wine. Over centuries of selective pressure, *O. oeni* has acquired and perfected various adaptive strategies that enable it to outcompete with other wine lactic acid bacteria during the later stages of vinification and thus to dominate in wine [15]. It proliferates in wine and cider during or after the yeast-driven alcoholic fermentation and reaches population levels above 10^6 cells/mL, thus becoming sometimes the only detectable bacterial species [24,25].

Wine pH is most selective, and, at a pH below 3.5, generally only strains of *O. oeni* can survive and express malolactic activity, while, in wines with a pH above 3.5, some *Lactobacillus* species have also shown a good ability to conduct MLF. Generally, the most frequent lactobacilli isolated from wine belongs to *Lb. plantarum*, *Lb. brevis*, *Lb. buchneri*, *Lb. hilgardii*, and *Lb. fructivorans*, although their occurrence and that of other species (i.e., *Lb. fermentum*, *Lb. kunkeei*, *Lb. mali*, *Lb. vini*) can be found depending on the grape varieties and typologies of wines [1]. Among them, *Lb. plantarum* is certainly the most important in wine because it is found frequently on grapes and in wine and is often involved in spontaneous MLF under high pH conditions. This versatile bacterium tolerates ethanol up to 14% (*v/v*) and can have similar SO_2 tolerance like *O. oeni*. Moreover, *Lb. plantarum* has a more diverse array of enzymes and can potentially exert positive effects on organoleptic properties of wine [1]. Some selected *L. plantarum* strains have shown interesting results for their capacity to induce MLF under high pH conditions, and, unlike *Oenococcus oeni*, *L. plantarum* has a facultative hetero-fermentative metabolism that prevents acetic acid production from hexose sugars. Due to these characteristics, selected strains of *Lb. plantarum* are currently being commercialized to induce MLF in wine [1,26].

Pediococcus damnosus is the other species well represented in the wine environment. It is often found after alcoholic fermentation in wines with rather high pH, along with *Lactobacillus* sp. and *O. oeni*. As it has been identified in most ropy wines, its presence is considered undesirable. In reality, only certain strains of *P. damnosus* are responsible for this spoilage and they are easily identified through polymerase chain reaction (PCR). Little research has been published on the possibility of using these organisms as wine LAB starter cultures. A study in which indigenous strains of *P. damnosus* dominate a starter culture of *O. oeni* and conduct the MLF shows that it is very capable of surviving in wine [27].

3. Selected Wine Lactic Acid Bacteria Starter Cultures and Wine Challenging Factors

Grape juice and (especially) grape wine contain a challenging matrix, with sugar, ethanol, organic acids, amino acids, fatty acids, other metabolites deriving from the yeast metabolism during alcoholic fermentation, phenol contents, pH, and SO_2 determining the growth of wine microorganisms. Various papers reported many factors that influence the occurrence of LAB and MLF in wines. Henick-Kling [28] and Wibowo et al. [29] listed, besides oxygen and CO_2, carbohydrates, amino acids, vitamins and minerals, organic acid content, the alcohol level, pH, and SO_2 level. The interrelationships between LAB and wine yeast [30] or other wine microorganisms and the method of vinification have been reported to be the most influential factors to affect LAB growth. The wine pH is one of the most important factors that limits LAB growth and MLF in wine [2,29,31] and determines the type of LAB which will be present. Ideally, for table wines, the pH should be between 3.1 and 3.6 [32], but due to global warming wine pH has increased in recent years in almost all wine regions.

3.1. Well-Known Factors that Affect Malolactic Fermentation and Bacteria Vitality

The best understood factors that govern successful MLF are SO_2, pH, alcohol, and temperature.

3.1.1. pH

The pH of the media has a drastic influence on the MLF itself as low pH inhibits the growth of the wine LAB. Most LAB are neutrophilic [33], and the optimum pH for the growth of lactic acid bacteria is close to neutrality. Some bacteria strains of the genera of *Lactobacillus* and *Oenococcus* show more acidophilic behavior. At pH values less than 3.0, bacterial growth is very difficult or impossible [34]. *Oenococcus oeni*, which is the organism of choice to conduct MLF under acidic conditions, will generally dominate, and in wine of pH above 3.5, strains of the genera of *Lactobacillus* or *Pediococcus* can be more present. The ability of the bacteria to obtain energy from the metabolism of glucose is inhibited at the low pH of wine [28]. *L. plantarum* shows a preference for malate as an energy source at low pH [35], even in the presence of glucose, which suggests this species as a starter for malate decarboxylation in fermenting musts [36]. Prahl [8] and Bou and Krieger [9]) proposed *Lactobacillus plantarum* as most promising for use as a starter culture in higher pH wines.

3.1.2. Ethanol

Ethanol is known for its bactericidal properties and it is the main yeast metabolite produced during alcoholic fermentation. It can play an integral role in the ability of wine LAB to survive in wine and induce the malolactic fermentation. It is difficult to specify the concentrations which will completely prevent LAB development. Radler [2], Peynaud and Domercq [3], and Henick-Kling [4] reported an increasing inhibition above 5% (*v/v*). Wibowo et al. [29] stated in their review that the ability of LAB to survive and grow in wine decreases as the alcohol concentration increases above 10% (*v/v*). Henick-Kling [6] indicated a strong impact of temperature on the toxicity of ethanol. A temperature of 25 °C and above, combined with alcohol levels above 14.5% (*v/v*) can inhibit bacterial growth and the malolactic fermentation. However, the ethanol tolerance is very strain dependent. Information in the literature is contradictory regarding the alcohol sensitivity for the various species of wine LAB. Davis et al. [5] reported strains of *Lactobacillus* and *Pediococcus* being in general more tolerant to high ethanol concentrations than *O. oeni*. From the observation of Wibowo et al. [29], most *Lactobacillus spp.* can tolerate about 15% (*v/v*). Britz and Tracey [37] acknowledged that all *O. oeni* strains are able to survive and grow at 10% (*v/v*) ethanol at pH 4.7. Studying the combined effects of pH, temperature, ethanol, and malate concentrations on *L. plantarum* and *O. oeni*. Guerzoni et al. [36] suggest *L. plantarum* being more competitive in early steps of alcoholic fermentation. However, more severe conditions, e.g., ethanol concentrations higher than 6% (*v/v*), favor *O. oeni*. Most robust commercial *O. oeni* strains, which are produced with a specific process for pre-adaptation (MBR™ process) to different wine conditions, show good survival and good malolactic activity up to 16% (*v/v*) alcohol, depending on other environmental factors even higher. With regard to commercial *L. plantarum* starter culture preparations for the induction of MLF in wine, there are two different approaches: Pre-inoculation: Prahl et al. [7] proposed inoculating must before alcoholic fermentation using a direct inoculum of a freeze-dried facultative hetero-fermentative *L. plantarum* starter culture. Inoculation before the wine yeast addition was recommended due to the sensitivity of the described *L. plantarum* strain towards alcohol. Contrastingly, a patent filed in 2004 [9] on "Alcohol-tolerant malolactic strains for the maturation of wines with average or high pH" relates to LAB strains of the genera *Lactobacillus* and *Pediococcus* displaying a good alcohol tolerance and the capability to induce a complete MLF upon direct inoculation into a wine with an alcohol content of 10% (*v/v*) or more and an average high pH level. More recently, more alcohol resistant *Lactobacillus* starter culture had been released, which can tolerate up to 15% (*v/v*) [10].

3.1.3. Temperature

In wine, the optimum temperature of growth is different from what is obtained in laboratory culture. The optimum range is dependent on other physical and chemical parameters of the wine, notably the ethanol content. A higher ethanol content will lead to a decrease in the optimum growth temperature. In general, MLF usually occurs at sub-optimal LAB temperatures (below or around 18 °C). At 15 °C or lower, the chance of bacterial growth is slight [38]. Guerzoni et al. [36] studied the effects of several chemico-physical factors (pH, SO_2, ethanol concentration and temperature) on *L. plantarum* and *O. oeni*. A temperature increase only positively affected the lag phase of *O. oeni*, but not of *L. plantarum*. A temperature increase exhibited a negative and positive influence on *O. oeni* and *L. plantarum*, respectively. The combination of high temperatures and high alcohol increase the toxicity of ethanol as outlined above. Low temperatures are not lethal but decrease the enzymatic activity.

3.1.4. Sulphur Doxide

Sulphur dioxide (SO_2) is another compound well known for its bactericidal action and plays an essential role in the growth of LAB and development of MLF [32]. This component is found in wine with variable concentrations according to the winemaking conditions and the yeast strain responsible for alcoholic fermentation. SO_2 is purposely added to wines to inhibit the growth of undesirable microorganisms and for its antioxidant effect. Sulphur dioxide in its free form, as well as in its bound form with aldehydes and ketones, is a potent inhibitor of many microbes, including LAB. Three liberated forms of SO_2 are present in wine: molecular SO_2, bisulphite (HSO_3^-), and sulphite (SO_3^{2-}). Molecular SO_2 is effective as a bacterial preservative [39], and a well-known synergistic effect is the impact of pH on the level of molecular SO_2. The lethal level of molecular SO_2 for most wine LAB is low (0.3 mg/L), but it is possible that certain selected wine LAB strains could have a better resistance to molecular SO_2. Depending on the pH of the juice/wine, the amount of molecular SO_2 is between 1% and 7% of the free SO_2 content. The molecular SO_2 increases with a decrease in pH and an increase in temperature and/or alcohol.

For MLF to be successful, the values of these chemical parameters described above must correspond to those that allow the bacterial cultures to function successfully. A favorable level of any of these components may compensate for an unfavorable level of one or several of the others. It is important to remember these factors function synergistically, i.e., their actions together have a greater total effect than the sum of their individual actions.

3.2. *Lesser-Known Factors that Affect Malolactic Fermentation*

A number of lesser known, but equally important, factors can influence the course of MLF and are outlined below.

3.2.1. Yeast Strain Selection

It has been known for some time that certain yeasts selected to conduct the alcoholic fermentation (AF) interact better with certain wine LAB for the successful achievement of MLF. Under specific conditions, certain yeast strains may produce high concentrations of SO_2, which has a negative influence on the growth and survival of the wine LAB. Similarly, yeast strains that exhibit an inordinate need for nutrients could exhaust the medium to such an extent that no reserve nutrients are available for the bacteria. Implementing a specific nutrition strategy for a particular yeast in the early stages of AF can largely surmount this [40–43] and avoid the production of certain unwanted metabolites or toxins derived from yeast stress. More recently, other bacterial growth inhibiters derived from yeast metabolism have been reported, such as medium-chain fatty acids [44] and yeast peptides (between 5 and 10 kDa) [45,46]. More recently, Liu et al. [30] reported certain peptides being stimulating for *O. oeni*. These effects depend on the nature and the level of fatty acids in the wine or the size of the yeast peptides, and can be exacerbated by low pH. On the other hand, the contact with the yeast lees has a very stimulating effect on MLF. The autolysis process releases amino acids and vitamins, and the must become richer in nutrients for the LAB. There may also be a detoxifying effect by yeast polysaccharides, as they may adsorb inhibitory compounds or complex them. Fumi et al. [10] reported yeast strains compatible with *O. oeni* starter cultures being also compatible with a *Lb. plantarum* starter culture strain.

3.2.2. Organic Acids

From practical experience, wines with L-malic acid levels below 1 g/L are not as conducive to MLF by *O. oeni*, as are wines with L-malic acid concentrations between 2 and 4 g/L. Wines with levels of L-malic acid above 5 g/L often start L-malic acid degradation, but do not go to completion. The cause is thought to be the result of the inhibition of the bacteria by increasing concentrations of L-lactic acid derived from the MLF itself. Since acidification with the organic acids lactic acid, L(−) or DL malic acid, L(+) tartaric acid and citric acid is authorized in many wine regions, and Vincent Gerbaux from the Institut Français de la Vigne et du Vin (IFV) in France has studied the influence of organic acid additions on the development of MLF (data not published). In this study, six selected wine LAB strains were inoculated into a Chardonnay wine, and five selected *O. oeni* wine LAB strains were inoculated into a Pinot noir wine, both of which were adjusted to a pH of 3.25. Increasing the amounts of L-malic acid, 0.75 to 5.2 g/L for Chardonnay and 3.0 to 5.7 g/L for Pinot noir, increased the time required to complete MLF, but the speed of L-malic acid degradation increased with increasing content of L-malic acid. The differences between selected wine LAB strains were observed. The addition of D-malic acid had no noticeable effect on MLF.

The presence of L-lactic acid in the wine inhibits the implantation and growth of the inoculated *O. oeni* strains, resulting in an inhibition of MLF. An initial content of L-lactic acid in the range of 1.5 g/L slows MLF, but a content of 3.0 g/L inhibits MLF by most of the tested *O. oeni* strains. Problems inducing MLF by inoculation with selected wine LAB cultures may be encountered when L-lactic acid was added to must or wine or in wines with a partial MLF. The impact of DL-lactic acid and D-lactic acid has yet to be investigated.

3.2.3. Tannins

Some red grape cultivars, such as Merlot, Tannat, and Zinfandel, may experience great difficulty undergoing successful MLF [47,48]. This may be related to certain grape tannins exerting a negative influence on the growth and survival of wine LAB, and consequently on the MLF. Research has been conducted exploring the impact of polyphenols on the growth and viability of wine LAB and their ability to degrade L-malic acid, often with inconsistent results. Polyphenols can have either stimulatory or inhibitory effects on the growth of wine LAB, depending on their type and concentration, and on the selected wine LAB strain in question. Figueiredo et al. [49], Chasseriaud et al. [50], and Stivala et al. [51] showed that tannin compositions containing a high percentage of condensed tannins can strongly affect the viability of *O. oeni* cells, whereas tannin blends consisting of anthocyanins and condensed tannins or catechin and epicatechin monomers and dimers, respectively, only slowed down the growth of bacteria when they were used at the highest concentration. These results are also in agreement with previous studies that showed no effect or a stimulatory effect of these compounds [49,52,53]. The successive activity of a 3-O-galloyl esterase and gallate decarboxylase, as it has been found in *L. plantarum* [54], may explain the stimulation by the addition of grape tannins.

3.2.4. Nutrient Deficiencies

In order to successfully complete MLF, proper nutrition for the wine LAB is of the utmost importance, because wine LAB are characterized as having complex nutritional requirements. Contrary to the fermentation yeast *Saccharomyces cerevisiae*, the bacteria *O. oeni* and other wine LAB cannot utilize inorganic nitrogen sources. Instead, sufficient amounts of organic nitrogen in the form of amino acids and peptides must be supplied. The vitamins pantothenic acid, thiamine, and biotin, as well as the trace elements Mg, Mn, and K, must also be provided to ensure bacterial growth and malolactic activity. Terrade and Orduña [55] investigated the essential growth requirements of four strains of wine LAB from the genera *Oenococcus* and *Lactobacillus*. The two *Oenococcus oeni* strains revealed a larger number of auxotrophies (18 and 23), the two *Lactobacillus* strains only had 11 and 14 auxotrophies. Despite the complex nutritional needs of wine LAB, the amounts they require are, in fact, quite small. Normally, the amount of nutrients contained in the wine matrix is sufficient to meet the needs of LAB, but certain vinification practices can result in nutrient deficiency.

3.2.5. Residual Lysozyme Activity and Chitosan Formulation

If lysozyme is used to control indigenous LAB during the production of wine, it is possible that residual levels of this enzyme may impact the duration of the subsequent MLF [56]. In most cases, racking the wine off the gross lees is recommended. Strains of *O. oeni* are more sensitive to the effects of lysozyme than strains of *Lactobacillus* or *Pediococcus* are.

Certain forms of chitosan, a natural polymer derived from chitin, exhibit antimicrobial properties. Chitosan is well known for its antimicrobial properties against yeast, bacteria, and fungi [57,58], and a preparation extracted from a fungal source has been used to neutralize contamination by *Brettanomyces bruxellensis* [59]. More recently, Chitosanglucan formulation has been released to inhibit wine lactic acid bacteria and delay or inhibit malolactic fermentation. To ensure timely MLF, as well as wine protection, the application of chitosan is recommended at the completion of MLF.

3.2.6. Others

Other inhibitory compounds include certain fungicides and pesticides, especially the former, which may have a detrimental effect on wine LAB. The residues of systemic fungicides are the inhibitoriest and they are often used in the later stage of grape maturation to control the fungus *Botrytis cinerea* [60]. More recently, mixes of medium-chain fatty acid or the addition of fumaric acid had been proposed to inhibit the growth of wine lactic acid bacteria and the resulting malolactic fermentation.

4. Selected Wine Lactic Acid Bacteria Starter Cultures and the Timing of Inoculation

Even when desirable malolactic acid bacteria are established in a winery, the onset of the MLF may take several months and may occur in some barrels and tanks but not in others. There are several options available to control MLF: firstly, the selection of conditions to encourage the growth of the indigenous malolactic flora; secondly, the induction of MLF in wines by inoculating with wines already undergoing MLF; or, thirdly, the induction of MLF by inoculation with either laboratory-prepared or commercial strains of LAB. Increased recognition of the influence of MLF on wine quality has led winemakers in recent years to seek better control over the occurrence and outcome of MLF. For this reason, the induction of the MLF by the use of selected LAB starter cultures is fast becoming the preferred option. Due to a massive inoculum with bacteria at 10^6 cfu/mL, less time is required for the bacteria to grow up to a cell density high enough to rapidly degrade the malate present in wine. Undesirable bacteria can be suppressed, which prevents wine alterations [1].

In the 1980s, commercially available strains of wine LAB from around the world became available to the wine industry. These LAB strains were selected from good spontaneous MLF and screened for good MLF kinetics, reliable performance under wine conditions, and desired sensory properties. They will tolerate the difficult growth and survival conditions found in wine and will produce compounds that impart positive sensory impacts to the wine. In the early 1990s, direct inoculation MLB starter cultures became available, and the most effective are pre-acclimatized during the production process. This step allows them to survive being added directly to wine, with no decrease in viable cell numbers, and no loss of malolactic activity. These wine LAB preparations can be added directly to wine or rehydrated in water for a short time prior to their addition.

Traditionally, when selected cultures of known wine LAB are used, inoculation is performed at the completion of AF. That said, already in 1985, Beelman and Kunkee [61] explored the possibility of inoculating wine LAB into juice along with the yeast used to conduct AF. Current thinking identifies the following times during wine production when selected wine LAB can be added (Figure 2).

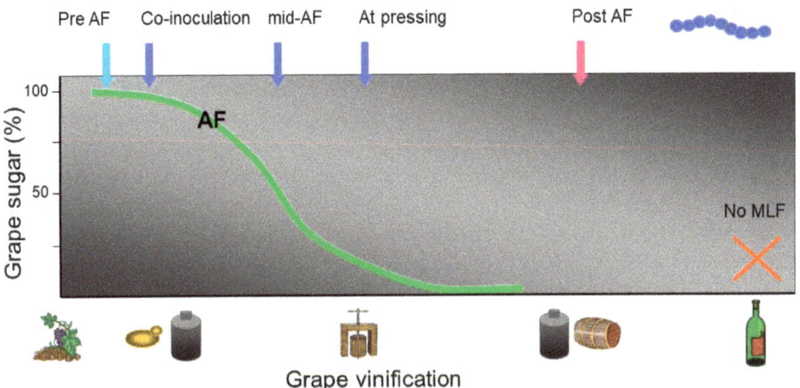

Figure 2. Inoculation regimes for selected wine lactic acid bacteria Adapted from Bartowsky, Australian Wine Research Institute (AWRI), Entretiens Scientifiques Lallemand Dubrovnik 2011.

4.1. *Co-inoculation with Selected Yeast and Selected Wine Lactic Acid Bacteria*

Before 2003, researchers at the Université de Bordeaux recommended making the wine LAB addition only after the completion of AF. They felt this timing would avoid the production of acetic acid and D-lactic acid, compounds derived from the hetero-fermentative carbohydrate metabolism of LAB [62]. They proposed that wine LAB added at earlier points during AF may result in slow or stuck yeast fermentation, or result in MLF inhibition due to yeast antagonism. To date, none of these concerns have been observed when both AF and MLF have been properly managed.

But other researchers proposed the inoculation of selected wine LAB into juice along with yeast because it was felt nutrient availability would be enhanced, and the absence of alcohol would allow wine LAB to better acclimatize to environmental conditions and grow more vigorously. Beelman and Kunkee [61] showed that MLF in the presence of fermentable sugars does not necessarily lead to the production of excessive amounts of acetic acid, as long as yeast fermentation starts promptly and goes to completion [63,64]. For a successful co-inoculation, some parameters are crucial for its success—choosing the right wine yeast, correctly rehydrated, good temperature management, and the proper yeast nutrition strategy are key points to integrate for any fermentation. Well-fed and healthy wine yeast and bacteria leads to complete and regular alcoholic and malolactic fermentations.

Since 2003, co-inoculation has gained increasing interest across all winemaking regions. Today, co-inoculation is understood as the practice of inoculating selected wine lactic acid bacteria at the beginning of the winemaking process shortly after yeast inoculation, usually 24 to 48 h after yeast inoculation. This technique is advantageous because not only will it secure the malolactic fermentation under most difficult conditions, but also because there are definite advantages that are recognized by winemakers and professionals: First bullet

- MLF can be completed in between 3 days and 2 weeks depending on the type of musts and the bacteria used.

- Co-inoculation to produce fresh wine styles with low diacetyl content: Co-inoculation always result in more fruit-driven wine styles and very low diacetyl content in wines. Early results also show that in the case of co-inoculation the high content of sugars could repress the metabolism of the diacetyl, as opposed to post-alcoholic fermentation inoculation. Moreover, under the reductive conditions generated by the active yeast, diacetyl produced will be immediately reduced to the less active metabolites, acetoin and butanediol.
- Co-inoculation to limit the development of *Brettanomyces* and off-flavors: The increase in sugar levels, pH, and sometimes lower SO_2 addition can influence the development of spoilage microorganisms, especially *Brettanomyces*, which can produce phenolic off-odors in wines. It is well known that the period from the end of AF to the start of MLF is particularly conducive to the development of *Brettanomyces*. Early inoculation with wine bacteria, either right after AF or in co-inoculation (24 h after inoculation with yeast), has proven to be a simple and effective method for preventing the development of *Brettanomyces* and the production of ethyl phenols off-flavors. Recent studies with IFV in Burgundy (Gerbaux) show co-inoculation with selected *O. oeni* strains can inhibit the growth of *Brettanomyces* (below 10 cell/mL) as opposed to the spontaneous control that is still contaminated with 500 cell/mL of *Brettanomyces* while the MLF is not completed and the wine is not stabilized [65].
- As a bio-control agent for low acidity/high pH wines, *Lactobacillus plantarum* with its facultative hetero-fermentative sugar metabolism is ideal as it completes MLF in 3–5 days during the alcoholic fermentation with no risk of increased volatile acidity due to its specific metabolism. It enables early stabilization of wines, as soon as the AF is finished.
- Co-inoculation as a tool for sustainability. In the frame of National Spanish R&D Project (VINySOST) involving six wineries, two companies of auxiliary industry, and several research centers (New strategies vine and winemaking for sustainable management in the production in great surfaces and increase in competitiveness of wineries in the international market—CDTI (strategic program CIEN, call 2014)), one of the studies were focused on the carbon footprint and analysis of life cycle from different axes involving wine producers. Within the study of carbon footprint and energy cost related to malolactic fermentation, co-inoculation with selected wine LAB had been compared to spontaneous MLF. Co-inoculated wine finished MLF with four after termination of the alcoholic fermentation whereas the spontaneous MLF took more than one and a half months to finish MLF. An electrical network analyzer was used after the energy consumption. Co-inoculation reduces the electricity consumption by more than 60%, as there was no need to heat the tanks to achieve a malolactic fermentation.

4.2. Sequential Inoculation with Selected Wine Lactic Acid Bacteria Post-alcoholic Fermentation

The traditional inoculation at the end of AF does not pose the risk of the bacterial decomposition of sugars and the resultant increase in VA, nor does the production of excessive amounts of lactic acid, known as "piqûre lactique," occur. Inoculation at this point avoids much of the toxicity attributed to some carboxylic acids, such as fumaric acid, as their concentration declines after AF [38]. The merit of inoculation at the end of AF can also be related to the availability of the bacterial nutrients, nitrogen-containing bases, peptides, amino acids, and vitamins that have arisen from yeast death and subsequent autolysis [66]. Another advantage may be simply from a logistical point of view. When using sequential inoculation, the wines that should undergo MLF can be separated from the wines where acidity is to be conserved. The vinification process can be conducted so that only one type of fermentation at a time is monitored. Often, this is perceived as less risk for cross-contamination.

However, exposure to high levels of ethanol that are present may result in delayed MLF, especially in wines produced in hot climates. If wine conditions are not limiting, selected wine LAB added after the AF are able to achieve cell concentrations comparable to those inoculated into must. In cases of nutrient limitation or adverse wine chemical parameters, the addition of a bacterial nutrient will support MLF. In instances where alcohol levels exceed 14.5% (*v/v*), selected wine LAB strains tolerant to alcohol must be used or they must be acclimatized before inoculating into wine. The additive, inhibiting effect of ethanol, pH, and SO_2 must be considered, and the strain best adapted to the conditions must be chosen.

5. Advantages of *Lactobacillus plantarum* Starter Cultures

Lactobacillus plantarum strains have shown most interesting results for their capacity to induce MLF under high pH conditions (Table 1), their facultative hetero-fermentative properties that avoid acetic acid production (Figure 3), and their more complex enzymatic profile compared to *O. oeni*, which could play an important role in the modification of wine aroma.

Table 1. Characteristics of Lactobacillus plantarum vs. Oenococcus oeni.

Species	*Lactobacillus plantarum*	*Oenococcus oeni*
Fermentation of sugars (hexoses)	Homo-fermentative = 2 × lactate	Hetero-fermentative = Lactate + acetate + CO_2
Wine parameter for best performance	pH > 3.5 Alcohol < 15.5% (*v/v*) Total SO_2 < 50 ppm Temperature 20–26 °C	pH > 3.1 Alcohol < 15.5% (*v/v*) Total SO_2 < 50 ppm Temperature > 17 °C
Genetic preposition for enzyme activities	MOST strains: Malolactic enzyme/ Glycosidase/Protease/Esterase/Lipase/Citrate lyase	Only very FEW strains: Malolactic enzyme Esterase/Protease/ Citratelyase/Methionine synthase c
Genetic preposition for bacteriocins production	Good potential	Only a FEW strains

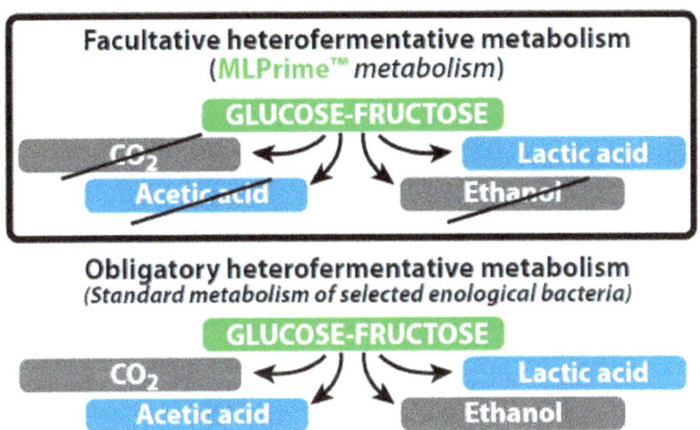

Figure 3. Sugar metabolism of wine lactic acid bacteria.

5.1. *Lactobacillus plantarum Starter Cultures for the Induction of Malolactic Fermentation in Must and Wine*

In 1988, the first malolactic starter culture was introduced to the wine industry. Prahl et al. [7] proposed to inoculate must before alcoholic fermentation using a facultative hetero-fermentative *L. plantarum* starter culture making use of non-proliferating cells. In EP 0398957B1, they disclosed a method of introducing a freeze-dried biomass of *L. plantarum* directly into must or fruit juice to induce MLF without significant consumption of sugars present in the must or fruit juice and substantially without any production of volatile acidity. The malolactic bacteria had been unable to survive in the fermented wine. The application under practical conditions asked for an inoculation 24 to 48 h prior to the addition of yeast with 10 g/hl of the freeze-dried preparation, corresponding to an inoculation level of 5×10^7 cells/mL wine [8]. Malic acid degradation had initiated rapidly; it slowed down and stopped when alcohol levels reached about 5–8% (*v/v*). At this stage, the *Lactobacillus* cultures died off. Depending on the wine parameters, mainly pH, temperature, and the speed of yeast fermentation, more or less malic acid is degraded. The advantage would have been partial malic acid degradation in low pH white wines. The disadvantages had been that microbial stability could not be achieved, since a part of the malic acid stayed in the wine as a source for the growth of other microorganisms. The amount of malic acid degraded is varying and not predictable. Furthermore, the application of this culture had only been recommended for low pH white wines, since above pH 3.5 the *L. plantarum* could grow and metabolize glucose producing L/D-lactic acid. This risk is quite high since lactic acid production would be significant and even yeast growth could be triggered by the excessive growth of this microorganism under high pH conditions. The use of this starter culture was limited, as the degree of malic acid conversion was variable and rarely complete, and due to the limited application in low pH wines only, along with the risk to leave the wines for two days without SO_2 and yeast addition.

In 2004, Bou and Krieger [9] filed a patent on an "Alcohol-tolerant malolactic strains for the maturation of wines with average or high pH". The patent had been published in 2004 under the international application number PCT/FR2004/001421. The patent relates to LAB strains of the genera *Lactobacillus* and *Pediococcus* capable of initiating and carrying out a complete MLF upon direct inoculation in dried, frozen, or lyophilized state, without a previous acclimatization step at a concentration of between 10^6–5×10^7 cfu/mL, into a wine with an alcohol content of 10% (*v/v*) or more and an average high pH level. The resistance to alcohol is apparent with an excellent survival rate on inoculation and a rapid start of malic acid degradation. Claims had been supported by examples of the successful induction of MLF under various wine conditions with *Lactobacillus* strains DSM-9916 and CNCM I-2924. These strains had been chosen from a pool of LAB strains selected for their good tolerance to various limiting conditions and specifically for high alcohol tolerance.

A more recent Italian selection led to *L. plantarum* strain V22 [10]. As part of a European project, where chemical adjuvant, wine yeast, and LAB were screened for their ability to degrade Ochratoxin A (OTA) in must and wine, three *L. plantarum* strains were selected at the University Catolica Sacro Cuore in Piacenza (UCSC) [67]. Ochratoxin A is a mycotoxin suspected of being nephrotoxic, teratogenic, hepatotoxic, and carcinogenic. *Lactobacillus plantarum* V22 showed the highest degradation of OTA under the experimental conditions. The three strains were tested in freeze-dried MBR® form for the induction of MLF in wine. Dried MLF starter cultures in MBR® form will allow direct inoculation into wine without significant loss of MLF activity. The *L. plantarum* strain V22 was the most robust under the tested conditions. This strain had been tested during three vintages under various high pH (>pH 3.5) and high alcohol conditions (≥14% (*v/v*)), proving to be as fast as *O. oeni* starter cultures when inoculated after alcoholic fermentation. Due to its facultative hetero-fermentative properties, *L. plantarum* is most interesting for co-inoculation without the risk of volatile acid formation when inoculated in the presence of sugars. Again, it proved to have a faster fermentation rate compared to *O. oeni*, if wine pH was higher than pH 3.5.

5.2. Specific Feature of *Lactobacillus plantarum* of Oenological Interest

Iorizzo et al. [68] selected 11 *L. plantarum* isolates from spontaneous MLF in wines from Southern Italy and characterized them according to their oenological characteristics and for their potential use as starter cultures for MLF in wine. None of the 11 strains produced biogenic amines which is an important criteria for its potential use as MLF starter culture. Cappozzi et al. [69] studied the biogenic amine degradation by *L plantarum* and found two strains able to degrade putrescine and tyramine under wine-like conditions.

Knoll et al. [70] studied LAB isolated from South African red wines during alcoholic and MLFs and 9 commercial malolactic bacteria starter strains including *L. plantarum* V22 for antimicrobial activity. Of the entire screened isolates, 26 strains, belonging to *L. plantarum*, *L. paracasei*, *L. hilgardii*, and *O. oeni*, showed activity towards various wine-related and non-wine-related indicator strains on a synthetic medium. A PCR-based screening revealed the presence of the plantaricin encoding genes plnA, plnEF, plnJ, and plnK in five selected *L. plantarum* strains, including V22. These strains have also been screened for bacteriocin activity by plate assays, on normal MRS media, MRS pH 3.5 and MRS 10% (*v/v*) ethanol (unpublished data). All 20 strains were tested against nine different sensitive organisms. Seven strains, including *L. plantarum* V22, showed bacteriocin inhibitory activity against all of the sensitive strains tested under those pH and ethanol conditions, but under real wine conditions bacteriocin producing activity was not expressed. Iorizzo et al. [68] could not detect a bacteriocin-producing activity within their selection of 11 *L. plantarum* strains from South Italian wines.

5.3. Mixed *Oenococcus oeni* and *Lactobacillus plantarum* Starter Cultures for the Induction of MLF

Lerm et al. [71] studied various *Oenococcus oeni* and *Lactobacillus plantarum* strains isolated from the South African wine environment for their potential use as malolactic starter cultures. These strains were characterized with regards to their properties of oenological interest, including the genetic screening for enzyme-encoding genes (enzymes implicated in wine aroma modification, as well as the absence of enzyme negatively impacting on of the final wine quality or integrity such as biogenic amine formation or production of ethylcarbamate), the ability to survive in wine, their fermentation capabilities, as well as their volatile acidity production. A total of three *O. oeni* and three *L. plantarum* strains were selected at the completion of this study. These strains showed the most potential during the characterization and were able to successfully complete MLF in Pinotage wine. It was again found that *L. plantarum* strains displayed a more diverse enzyme profile than *O. oeni* strains, particularly with regards to the presence of the aroma-modifying enzymes β-glucosidase and phenolic acid decarboxylase (PAD), which implies the future use of this species in the modification of the wine aroma profile and use as commercial starter culture. It was concluded that *Lactobacillus plantarum* strains might have an added beneficial influence in that it has the genetic potential to influence the wine aroma profile to a larger extent than *O. oeni*, due to its cache of enzymes. Based on outcome of this study a mixed starter culture consisting of an *Oenococcus oeni* and a *Lactobacillus plantarum* strain has been introduced 2011 as "Co-Inoculant" for simultaneous inoculation together with the yeast for induction of malolactic fermentation. This was the first commercially available blend of its kind in the world recommended for co-inoculation in high pH grape musts (>pH 3.4) only. Nowadays, a second blend *O. oeni/L. plantarum* is on the market, which can work at lower pH (>pH 3.2).

5.4. A New Concept of *Lactobacillus plantarum* Starter Cultures for High pH Red Wines.

Although co-inoculation (inoculation of selected wine lactic acid bacteria 24 to 48 h after the inoculation with selected wine yeast) is getting very popular and is more and more applied because of its various benefits outlined in chapter 4.1, some winemakers still consider co-inoculation with *O. oeni* as risky because of their obligatory hetero-fermentative properties. They wrongly fear co-inoculation, although this practice has more than proven itself to be a secure choice for high pH red wines (above pH 3.5) in which the native flora is even more critical. The biggest fear is to get a stuck alcoholic fermentation due to antagonism with the wine LAB, and having the bacteria taking over, resulting in an important increase in volatile acidity due to the hetero-fermentative metabolism of the residual sugars. More recently, a new starter culture called ML Prime™ (a pure *Lactobacillus plantarum* ferment) was released. Due to its specific optimized production process, this *Lactobacillus plantarum* starter culture expresses extremely high malolactic enzymatic activity as soon as it is added to must. MLF is therefore completed in record time (3–7 days in average) during alcoholic fermentation (Figures 4 and 5), and, unlike classical inoculum with *O. oeni*, no further growth is needed, which explains the very rapid onset of MLF upon inoculation into the must without any impact on yeast vitality and alcoholic fermentation. As explained before, *L. plantarum* degrades hexose sugars only via the homo-fermentative pathway, so there is no risk of acetic acid production from residual sugars that may be present in high pH wines, or still present, when MLF has finished before the end of the alcoholic fermentation.

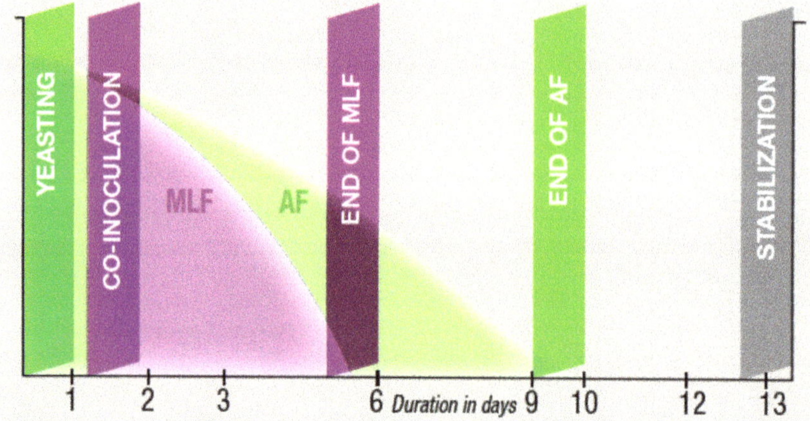

Figure 4. Vinification with a new generation starter culture of *Lactobacillus plantarum*.

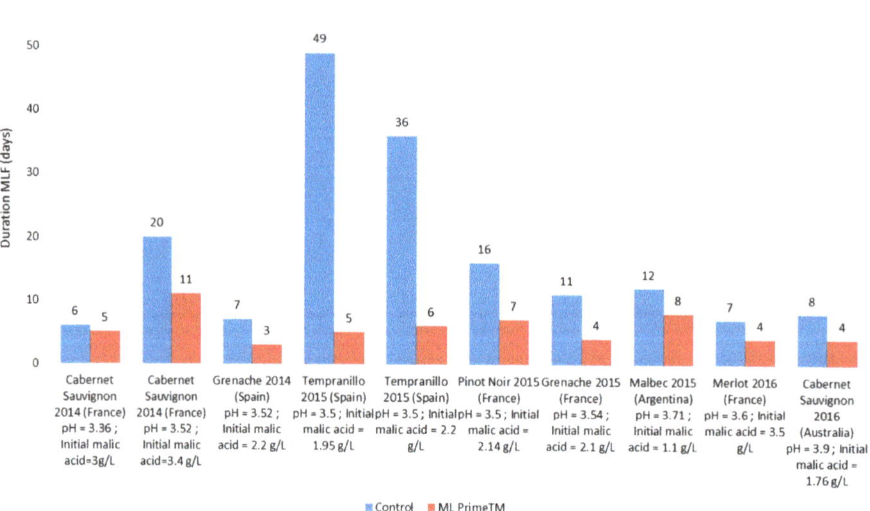

Figure 5. Duration of Malolactic fermentation (MLF) in various trials (days) inoculated with *L. plantarum* ML-PRIME™ (internal data).

5.4.1. Control of Microbial Contamination

As a result of the global warming, wines with a pH of over 3.5–3.7 are more and more frequent. At those pH levels, we can observe very fast growth of various indigenous microorganisms, some of which are spoilage bacteria that can cause a loss of wine quality or present health concerns. Co-inoculation is advantageous because it allows for the early dominance of a selected wine LAB strain and the faster onset and completion of MLF and early wine stabilization [42].

A more recent OIV regulation (OIV-Oeno-264-2014) on good vinicultural practices for controlling *Brettanomyces* proposed the co-inoculation of selected yeast and selected wine bacteria to shorten the lag phase between the end of alcoholic fermentation and the start of malolactic fermentation and consequently limit the implantation and the growth of *Brettanomyces*, another wine spoilage microorganism with a detrimental impact on wine quality. It also states that the use of malolactic starters is a good way to limit the development of *Brettanomyces* and the production of the undesired compounds

4-ethylphenol, 4-ethylguaïacol, and 4-ethylcatechol. These volatile phenols are characterized by animal-like off-flavors described as horse and barnyard, and/or pharmaceutical odors characterized as medicinal. Because of the high initial vitality of *L. plantarum* ML Prime™, an immediate onset of malolactic fermentation can be observed, as evidenced in Figure 4 where malic acid was degraded during alcoholic fermentation. However, it is important to respect the windows of application summarized in Table 2, which is narrower than for an *O. oeni* starter culture.

Table 2. Optimum conditions for the use of *Lactobacillus plantarum* ML Prime™.

Types of Wines	**Reds–Traditional Vinification (Short or Medium Maceration–Thermovinification (Liquid Phase) Initial Sugar/Potential Alcohol up to 260 g/L/15,5% (*v*/*v*))**
Timing of Bacteria Inoculation	Only co-inoculation Addition of ML Prime™, 24 h after adding yeast
SO2 Addition on Grapes/Must	≤5 g/hL
pH **Acid Malic Content**	≥ 3.4 maximum 3 g/L
Temperature During AF	20° to 26 °C

5.4.2. Malolactic Fermentation and Red Wine (Pinot Noir) Color

In Pinot Noir, MLF is often delayed because the resulting wines have anecdotally been reported to have superior color. Delayed MLF in a Pinot noir wine, for up to 4 months, showed improved wine color intensity [72]. Pinot noir wine color presents its own unique challenges, particularly because of its low tannin and anthocyanin content, with a bias towards the less stable acetylated form. The formation of wine color is a complex reaction with many different factors having and integral role. It is known that microbial metabolites, acetaldehyde, and pyruvic acid play a role in the formation of polymeric pigments [73]; however, the degradation of these compounds by *O. oeni* and the impact it could have on red wine color is not been well understood. A study in Pinot noir wine showed that there was no significant impact on color loss (A520) when MLF was delayed by up to six months; however, there was an impact on the formation of polymeric pigments [74] This study demonstrated the role of acetaldehyde and/or pyruvic acid degradation by *O. oeni* during MLF as a cause for reduced polymeric pigment formation independent of the pH change.

The results from a research collaboration with the Oregon State University and the team of James Osborn [75] showed different LAB species and strains can metabolize acetaldehyde at different rates (Table 3), which then in turn will affect red wine color post MLF (Figure 6). *Lactobacillus plantarum* (ML-Prime™) metabolizes acetaldehyde at a slower rate to *O. oeni* strains (*O. oeni* OM and *O. oeni* AL).

Table 3. Acetaldehyde concentration (mg/L) in a Pinot noir wine pre- and post-MLF and a control wine without MLF extracted from Bartowsky and Krieger-Weber [75].

	Pre-MLF	**Post-MLF**
Control	8.37	9.2
***L. plantarum* ML-Prime™**	8.37	7.67
***O. oeni* OM**	8.37	2.33
***O. oeni* AL**	8.37	1.7

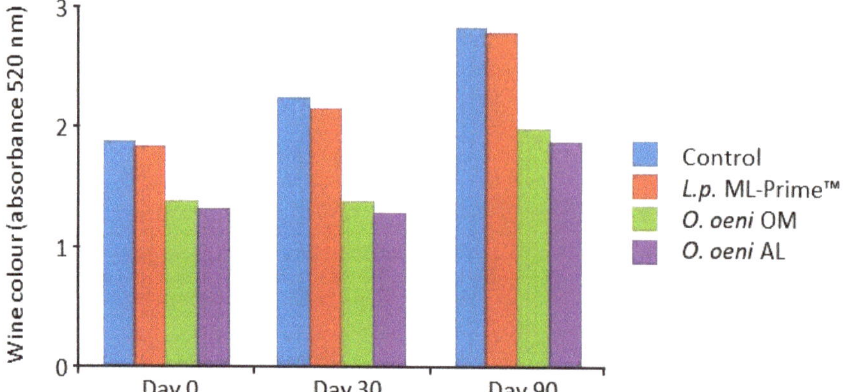

Figure 6. Wine color (520 nm) of Pinot noir wine that did not undergo MLF (Control) or MLF with different malolactic bacteria strains; the wines were stored at 13 °C and extracted from Bartowsky and Krieger-Weber [75].

When color and polymeric pigment values were measured in the wines post-MLF (Day 0) and after 30 and 90 days storage at cellar temperatures (Figure 6), a reduction in color was observed in wines that underwent MLF with *O. oeni* AL or *O. oeni* OM, whereas less loss of color was noted in wines that underwent MLF with the *L. plantarum* (ML-Prime™). After 30 or 90 days of aging, no loss of polymeric pigment was noted in wines that underwent MLF with ML-Prime™.

The overall color of Pinot noir wine can be better managed by selecting a specific wine lactic acid bacteria with consideration of the acetaldehyde metabolism and timing of MLF inoculation. Delaying MLF can actually also promote the combination of tannins and anthocyanins, resulting in a lesser impact of SO_2 on color. However, this approach to use a delayed MLF for more and stable color must be carefully weighed up against potential microbial spoilage, including Brettanomyces and biogenic amine formation (indigenous LAB).

5.5. A New Concept of Lactobacillus plantarum Starter Cultures for Low pH White Wines

Although it was out of the comfort zone for *L. plantarum*, ML-Prime (pH ≥ 3.4 and malic acid maximum 3 g/L) was also tested in white wines, and the results had been surprisingly good. Due to the optimized production process resulting in a very high de-acidification activity, it showed an excellent performance when added into the must (24 h after the yeast) or wine with initial low pH and high malic acid content. Even at a pH as low as 3.05 it allows a partial degradation of the malic acid in the white wine vinification process. The percentage of malic acid degradation depends on the specific must or wine conditions (pH, acid malic content, the total acidity, temperature, and the SO_2 content) and the grapes varietals, and can vary between 20% and 90%. Figure 7 shows the kinetics of malic acid degradation with *L. plantarum* (ML-Prime™) in a 2017 Chardonnay from South of France with an initial malic acid concentration of 3.6 g/L adjusted to pH 3.1, pH 3.2, pH 3.3, pH 3.4, or pH 3.5, respectively. Chardonnay is known for its difficulties to undergo MLF. *L. plantarum* was able to degrade about 30% of the malic acid (between 1.1 and 1.5 g/L malic acid had been degraded) when the wine pH ranged from pH 3.1 to pH 3.5, whereas, at pH 3.5, a complete malic acid degradation was achieved.

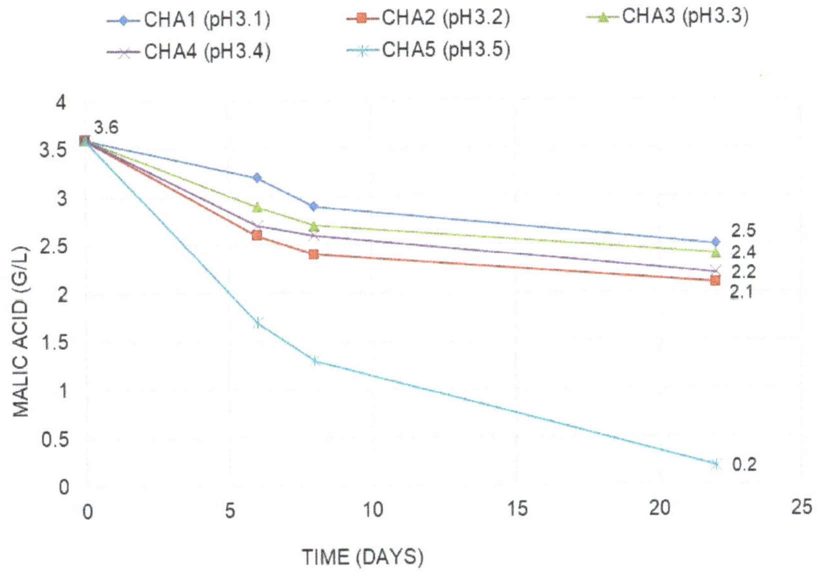

Figure 7. Kinetics of malic acid degradation in a 2017 Chardonnay (South-France) after co-inoculation with *L. plantarum* (ML-Prime™) depending on the pH. Initial must analyses: malic acid concentration 3.6 g/L, total sugars 189 g/L, potential alcohol 11.2% (*v/v*), 7.56 g/L total acidity (in tartaric acid), nitrogen 193 mg/L (Internal laboratory trials Lallemand SAS).

As the wine matrix is very versatile, a precise prediction of how much malic acid will remain in the wine is not possible. Figure 8 shows the malic acid degradation in sequential inoculation in a 2019 Chardonnay from Germany. Again, the matrix was a difficult Chardonnay with a pH of 3.21, 13.2% (*v/v*) alcohol, and 55 mg/L total SO_2, the temperature was at 17 °C. Under these challenging conditions, most of the *O. oeni* strains failed in sequential inoculation. Only one *O. oeni* strain OM started malolactic fermentation, but only after re-inoculation and with a very slow degradation. The *L. plantarum* culture (ML-Prime™) finished MLF when inoculated with the normal inoculation ratio after 25 days, and, when doubling the inoculation dosage MLF, it was finished within 10 days. In this experiment, the *L. plantarum* cultures maintained a high cell viability throughout the malolactic fermentation. Contrastingly, this Chardonnay wine must have contained a toxic compound, which had trigged *O. oeni*, as upon inoculation a sharp die-off of the *O. oeni* population could be observed. Only *O. oeni* Oo OM showed better survival and could finally implant, regrow, and induce MLF in this wine.

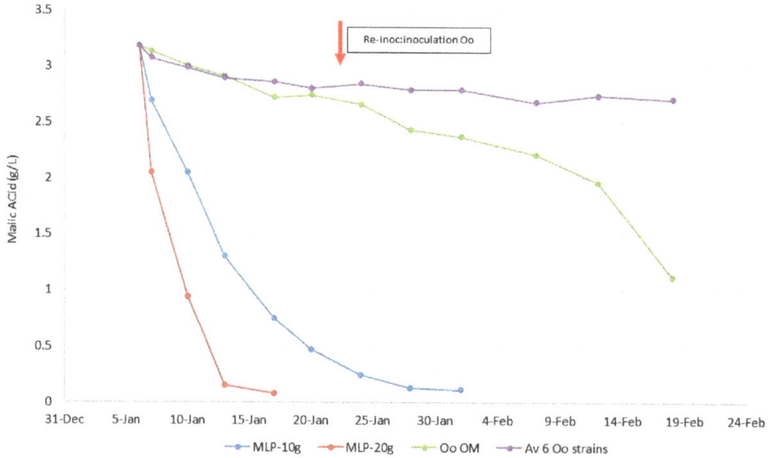

Figure 8. Kinetics of malic acid degradation in a 2019 Chardonnay (Germany) after sequential inoculation with *L. plantarum* (ML-Prime™), and different *O. oeni* strains. The kinetics show malic acid degradation with the *L. plantarum* culture inoculated with a single and double dosage and *O. oeni* strain Oo OM, and the average of 6 other *O. oeni* strains inoculated at a single dosage and re-inoculated at a double dosage. Wine analyses before inoculation: malic acid concentration 3.18 g/L, pH 3.21, 13.2% (*v/v*), 55 mg/L total SO_2 (Internal data).

Knowing the specifications above, it can be a tool for winemakers who want to achieve only a partial MLF in case of white wines vinification or a biological de-acidification instead of chemical de-acidification. The window of application for *L. plantarum* ML-Prime™ is outlined below:

- pH: ≥3.05
- Malic acid content: ≤8 g/L
- Temperature range: from 17 °C to 22 °C
- Total SO_2 tolerance in must up to 5 g/hL
- Free SO_2 tolerance in wines: less than 10 mg/L

5.6. Interesting Sensory Properties of Lactobacillus plantarum in Wine Application

As outlined by Lerm et al. [71], *Lactobacillus plantarum* strains might have an added beneficial influence, as it has the genetic potential to influence the wine aroma profile to a larger extent than *O. oeni*, due to its cache of enzymes. Mtshali et al. [76] conducted a genetic screening for wine-related enzymes within *Lactobacillus* species isolated from South African wines. They found a range of genes encoding for b-glucosidase, protease, esterase, citrate lyase (a-, b- and c-subunits), and phenolic acid decarboxylase. These findings indicated a possible use of *L. plantarum* not only for conducting MLF but also as the potential source of enzymes to impact positively on wine aroma, but expression under wine conditions needs further investigation. The commercial starter strain *L. plantarum* strain V22 had been included in a genetic screening of winemaking LAB starter strains mainly belonging to the species *O. oeni* for wine-relevant enzymatic activities [11]. The enzymes of interest that were screened for included β- glucosidase, esterase, protease, and phenolic acid decarboxylase (PAD). The V22 strain was found to possess more diverse enzymatic profiles related to aroma than *O. oeni*. The biggest differences were observed for the presence of esterase, protease, and PAD. The findings of Iorizzo et al. [68] reported the release of free volatiles from odorless glyosidic aroma precursors by all 11 *L. plantarum* strains in their study in a synthetic wine medium. Interestingly, the *L. plantarum* strain M10 was not only a major producer of 1-octanol, but also released a considerable amount of other odorant compounds with low odor thresholds. Still, these findings need to be validated in a real wine matrix. Spano et al. [77] reported that the expression of β-glucosidase gene in *L. plantarum* is regulated by abiotic stresses such as ethanol, temperature, and pH.

Still, the application of this species in grape must and wine is rather new since only recently commercial starter cultures had been made available, which can survive also at higher alcohol levels and can induce a reliable malolactic fermentation in wine. Further research is needed to elaborate the sensory contribution of these species to the wine aroma profile.

5.7. Other Applications of *Lactobacillus plantarum* Apart from the Induction of Malolactic Fermentation

However, the application of *L. plantarum* in vinification should not be restricted to de-acidification through malolactic fermentation only, Lucio et al. [78,79] have most recently proposed *L. plantarum* for the biological acidification of wines. Within the project CENIT CDTI 2008, they selected *Lactobacillus* strains, which show a high potential as biological acidification starters for winemaking when inoculated prior to the alcoholic fermentation into high pH grape must. WO 2015/110484A2 patent application [80] proposes reverse inoculation (inoculation prior to the inoculation with selected wine yeast) or the co-inoculation (inoculation together with the wine yeast) of selected homo-fermentative or facultative hetero-fermentative lactic acid bacteria strains to produce fermented fruit beverages, such as wine or cider, with a reduced alcohol level. Moreover, the international patent application WO 2015/110484 A2 [81] relates to the use of lactic acid bacteria as bio-protective agents against unwanted microorganisms, such as mold and gram-negative bacteria, such as acetic acid bacteria. The inventors propose a specific *L. plantarum* strain as an antimicrobial agent in the process of winemaking.

6. Conclusions

Lactobacillus is one of the most diverse group of microorganisms associated with the wine environment. Some *Lactobacillus* species have also displayed the ability to survive the harsh wine conditions, and, within this group, the species *Lactobacillus plantarum* has shown the most potential as a starter culture for the induction of malolactic fermentation. Global warming and the trend towards harvesting higher maturity grapes have resulted in the processing of higher pH musts. Microbial stability as a result of lactic acid bacteria activity will play a more important role in the warmer climates. Under these high pH conditions, *Lactobacillus plantarum* bacteria have shown especially interesting results, not only for their capacity to induce malolactic fermentation when inoculated either shortly after the yeast (co-inoculation) into the must or in sequential inoculation after alcoholic fermentation, but also for their homo-fermentative properties for the metabolism of hexose sugars, which minimizes the risk of acetic acid production. *L. plantarum* was found to dispose over a more complex enzymatic system, which could play role in the modification of wine aroma. More research is certainly needed to study the expression of these enzyme activities in wine.

Applying a strong *Lactobacillus plantarum* inoculum with high malolactic activity assures the early onset of predictable and complete MLF in a short period of time (during AF) and allows an early stabilization of the wine. Even under limiting white wine conditions, a complete or partial malolactic fermentation can be induced. Since this species is very versatile, other application for bio-protection and acidification mat play a more important role in the use of this starter culture.

Funding: This research received no external funding.

Acknowledgments: The authors thank research partner James Osborne (Oregon State University USA) for the research collaboration on impact of malolactic fermentation on Pinot Noir color, and our colleague Magali Déléris-Bou from our research Laboratory at Lallemand SAS in Toulouse, for conducting the internal experiments.

Conflicts of Interest: The authors declare no conflict of interest.

References

1. Du Toit, M.; Engelbrecht, L.; Lerm, E.; Krieger-Weber, S. *Lactobacillus*: The next generation of malolactic fermentation starter cultures-an overview. *Food Bioprocess Technol.* **2011**, *4*, 876–906. [CrossRef]
2. Radler, F. Die mikrobiologischen Grundlagen des Säureabbaus in Wein. *Zentralbl Bakteriol Parasitenkd* **1966**, *120*, 237–287.
3. Peynaud, E.; Domercq, S. Étude de quatre cents souches de coques hétérolactiques isolées de vins. *Ann. Inst. Pasteur* **1968**, *19*, 159–170.
4. Henick-Kling, T. Growth and Metabolism of *Leuconostoc oenos* and *Lactobacillus plantarum* in Wine. Ph.D. Thesis, University of Adelaide, Adelaide, Australia, 1986.
5. Davis, C.R.; Wibowo, D.; Fleet, G.H.; Lee, T.H. Properties of wine lactic acid bacteria: Their potential enological significance. *Am. J. Enol. Vitic.* **1988**, *39*, 137–142.
6. Henick-Kling, T. Malolactic Fermentation. In *Wine Microbiology and Biotechnology*, 1st ed.; Fleet, G.H., Ed.; Harwood Academic Publishers: Chur, Switzerland, 1993; pp. 289–326.
7. Prahl, C. Method of Inducing the Decarboxylation of Malic Acid in Must or Fruit Juice. European Patent International Application Number PCT/DK89/00009, 24 January 1989.
8. Prahl, C. La décarboxylation de l'acide L-malique dans le moût par l'ensemencement de lactobacilles homofermentaires. *Revue des Œnologues* **1989**, *54*, 13–17.
9. Bou, M.; Krieger, S. Alcohol-Tolerant Malolactic Strains for the Maturation of Wines with Average or High pH. U.S. Patent 7,625,745, 9 June 2004.
10. Fumi, M.D.; Krieger-Weber, S.; Déléris-Bou, M.; Silva, A.; du Toit, M. A New Generation of Malolactic Starter Cultures for High pH Wines. WB3 Microorganisms—Malolactic-Fermentation. In Proceedings of the International IVIF Congress, Düsseldorf, Germany, 21–23 March 2010.
11. Soerensen, K.; Kibenich, A.; Johansen, E. Lactobacillis planaturm Ceels with Improved Resistance to High Concentrations of Ethanol. International Patent Application Number PCT/EP2012/061296, 14 June 2012.
12. Krieger-Weber, S.; Silvano, A.; Déléris-Bou, M.; Vidal, F.; Dumont, A. Neues Konzept für die Mikroflora. *Das Deutsche Weinmagazin* **2017**, *19*, 30–32.
13. Lerm, E. The Selection and Characterization of Lactic Acid Bacteria to be Used as a Mixed Starter Culture for Malolactic Fermentation. Master's Thesis, Agricultural Science, Stellenbosch University, Stellenbosch, South Africa, 2010.
14. Bae, S.; Fleet, G.H.; Heards, G.M. Lactic acid bacteria associated with wine grapes from several Australian vineyards. *J. Appl. Microbiol.* **2006**, *100*, 712–727. [CrossRef]
15. Ribéreau-Gayon, P.; Dubourdieu, D.; Donèche, B.; Lonvaud, A. Handbook of Enology, Volume 1. In *The Microbiology of Wine and Vinifications*; Ribéreau-Gayon, P., Ed.; Wiley: Chichester, UK, 2006; pp. 149–151.
16. Fugelsang, K.C.; Edwards, C.G. *Wine Microbiology: Practical Applications and Procedures*; Fugelsang, K.C., Edwards, C.G., Eds.; Springer: New York, NY, USA, 1997.
17. Jackson, R.S. Origin and Growth of Lactic Acid Bacteria. In *Wine Science: Principles and Applications*; Jackson, R.S., Ed.; Academic Press: Cambridge, MA, USA, 2008; p. 394.
18. Boulton, R.B.; Singleton, V.L.; Bisson, L.F.; Kunkee, R.E. *Principles and Practices of Winemaking*; Boulton, R.B., Ed.; Chapman and Hall Publishers: New York, NY, USA, 1996.
19. Dicks, L.M.T.; Endo, A. Taxonomic status of lactic acid bacteria in wine and key characteristics to differentiate species. *S. Afr. J. Enol. Vitic.* **2009**, *30*, 72–90. [CrossRef]
20. Volschenk, H.; van Vuuren, H.J.J.; Viljoen-Bloom, M. Malic acid in wine: Origin, function and metabolism during vinification. *S. Afr. J. Enol. Vitic.* **2006**, *27*, 123–136. [CrossRef]
21. Berbegal, C.; Peña, N.; Russo, P.; Grieco, F.; Pardo, I.; Ferrer, S.; Spano, G.; Capozzi, V. Technological properties of *Lactobacillus plantarum* isolated from grape must fermentations. *Food Microbiol.* **2016**, *57*, 187–194. [CrossRef]
22. Valdés La Hens, D.; Bravo-Ferrada, B.M.; Delfederico, L.; Caballero, A.C.; Semorile, L.C. Prevalence of *Lactobacillus plantarum* and *Oenococcus oeni* during spontaneous malolactic fermentation in Patagonian wines revealed by polymerase chain reaction-denaturing gradient gel electrophoresis with two targeted genes. *Aust. J. Grape Wine Res.* **2015**, *21*, 49–56. [CrossRef]

23. López-Seijas, J.; García-Fraga, B.; Da Silva, A.F.; Zas-García, X.; Lois, L.C.; Gago-Martínes, A.; Leão-Martins, J.M.; Sieiro, C. Evaluation of malolactic bacteria associated with wines from Albariño variety as potential; starter cultures for quality and safety. *Foods* **2020**, *9*, 99. [CrossRef]
24. Fleet, G.H.; Lafon-Lafourcade, S.; Ribereau-Gayon, P. Evolution of yeasts and lactic acid bacteria during fermentation and storage of Bordeaux wines. *Appl. Environ. Microbiol.* **1984**, *48*, 1034–1038. [CrossRef] [PubMed]
25. Lonvaud-Funel, A. Lactic acid bacteria in the quality improvement and depreciation of wine. *Antonie Van Leeuwenhoek* **1999**, *76*, 317–331. [CrossRef] [PubMed]
26. Bartowky, E.J.; Costello, P.J.; Chambers, P.J. Emerging trends in the application of malolactic fermentation. *Aust. J. Grape Wine Res.* **2015**, *21*, 663–669. [CrossRef]
27. Juega, M.; Costantini, A.; Bonello, F.; Cravero, M.C.; Martinez-Rodriguez, A.J.; Carrascosa, A.V.; Garcia-Moruno, E. Effect of malolactic fermentation by *Pediococcus damnosus* on the composition and sensory profile of Albariño and Caiño white wines. *J. Appl. Microbiol.* **2013**, *116*, 586–595. [CrossRef]
28. Henick-Kling, T. *Yeast and Bacteria Control in Winemaking, Modern Methods of Plant Analysis*; Linskens, R.S., Jackson, J.F., Eds.; Springer: Berlin/Heidelberg, Germany, 1988; Volume 6, pp. 276–315.
29. Wibowo, D.; Eschenbruch, R.; Davis, D.R.; Fleet, G.H.; Lee, T.H. Occurrence and growth of lactic acid bacteria in wine: A review. *Am. J. Enol. Vitic.* **1985**, *36*, 302–313.
30. Liu, Y.; Forcisi, S.; Harir, M.; Deleris-Bou, M.; Krieger-Weber, S.; Lucio, M.; Longin, C.; Degueurce, C.; Gougeon, R.D.; Schmitt-Kopplin, P.; et al. New molecular evidence of wine yeast-bacteria interaction unraveled by non-targeted exometabolomic profiling. *Metabolomics* **2016**, *12*, 1–16. [CrossRef]
31. Dittrich, H.H.; Grossmann, M. Der Mirobielle Säureabbau. In *Mikrobiologie des Weines, Handbuch der Getränketechnologie*, 3rd ed.; Dittrich, H.H., Grossmann, M., Eds.; Ulmer Verlag: Stuttgart, Germany, 2005; pp. 151–169.
32. Amerine, M.A.; Berg, H.W.; Kunkee, R.E.; Ough, C.S.; Singleton, V.L.; Webb, A.D. The Composition of Grapes and Wines. In *The Technology of Wine Making*, 4th ed.; Amerine, M.A., Berg, H.W., Eds.; Avi Publishing Company: Westport, CT, USA, 1980.
33. Nicolas, D. Caractérisation Physiologique et Moléculaire de Préparations Malolactiques de *Oenococcus oeni* Destinées à L'ensemencement Direct des Vins. Ph.D. Thesis, University of Dijon, Dijon, France, 2005.
34. Lonvaud-Funel, A. Microbiology of the malolactic fermentation: Molecular aspects. *FEMS Microbiol. Lett.* **1995**, *126*, 209–214. [CrossRef]
35. Cox, D.J.; Henick-Kling, T. Chemiosmotic energy from malolactic fermentation. *J. Bacteriol.* **1989**, *171*, 5750–5752. [CrossRef]
36. Guerzoni, M.E.; Sinigaglia, M.; Gardini, F.; Ferruzzi, M.; Torriani, S. Effects of pH, temperature, ethanol, and malate concentration on *Lactobacillus plantarum* and *Leuconostoc oenos*: Modelling of the malolactic activity. *Am. J. Enol. Vitic.* **1995**, *3*, 368–374.
37. Britz, T.J.; Tracey, R.P. The combination effect of pH, SO_2, ethanol and temperature on the growth of *Leuconostoc oenos*. *J Appl. Bacteriol.* **1990**, *68*, 23–31. [CrossRef]
38. Lafon-Lafourcade, S. Chapter III: Lactic Acid Bacteria and the Malolactic Fermentation. In *Wine and Brandy in Biotechnology*; Rehm, H.J., Redd, G., Eds.; Verlag Chemie: Weinheim, Germany, 1983; pp. 81–163.
39. Chang, I.S.; Kim, B.H.; Shin, P.K. Use of sulfite and hydrogen peroxide to control bacterial contamination in ethanol fermentation. *Appl. Environ. Microbiol.* **1997**, *63*, 1–6. [CrossRef] [PubMed]
40. Massera, A.; Soria, A.; Catania, A.; Krieger, S.; Combina, M. Simultaneous Inoculation of Malbec (*Vitis vinifera*) Musts with Yeast and Bacteria: Effects on Fermentation Performance, Sensory and Sanitary Attributes of Wines. *Food Technol. Biotechnol.* **2009**, *47*, 192–201.
41. Abrahamse, C.; Bartowsky, E. Inoculation for MLF reduces overall vinification time. *Aust. N. Z. Grapegrow. Winemak.* **2012**, *577*, 41–46.
42. Azzolini, M.; Tosi, E.; Vagnoli, P.; Krieger, S.; Zapparoli, G. Evaluation of technological effects of yeast-bacterial co-inoculation in red table wine production. *Ital. J. Food Sci.* **2010**, *3*, 257–263.
43. Jussier, D.; Dubé Morneau, A.; Mira de Orduña, R. Effect of simultaneous inoculation with yeast and bacteria on fermentation kinetics and key wine parameters of cool-climate Chardonnay. *Appl. Environ. Microbiol.* **2006**, *72*, 221–227. [CrossRef]
44. Renouf, V.; La Guerche, S.; Moine, V.; Murat, M.L.; Bowyer, P.K. Malolactic fermentability of wines: The role of octanoic and decanoic acid. *Aust. N. Z. Grapegrow. Winemak.* **2010**, *560*, 93–98.

45. Osborne, J.P.; Charles, G.E. Inhibition of malolactic fermentation by a peptide produced by *Saccharomyces cerevisiae* during alcoholic fermentation. *Int. J. Food Microbiol.* **2007**, *118*, 27–34. [CrossRef]
46. Nehme, N.; Mathieu, F.; Taillandier, P. Impact of the co-culture of *Saccharomyces cerevisiae—Oenococcus oeni* on malolactic fermentation and partial characterization of a yeast-derived inhibitory peptidic fraction. *Food Microbiol.* **2010**, *27*, 150–157. [CrossRef]
47. Lonvaud-Funel, A. Interactions between Lactic Acid Bacteria of Wine and Phenolic Compounds. In *Nutritional Aspects II, Synergy between Yeast and Bacteria*; Lallemand Technical Meeting: Perugia, Italy, 2001; pp. 27–32.
48. Vivas, N.; Augustin, M.; Lonvaud-Funel, A. Influence of oak wood and grape tannins on the lactic acid bacterium *Oenococcus oeni* (*Leuconostoc oenos*). *J. Sci. Food Agric.* **2000**, *80*, 1675–1678. [CrossRef]
49. Figueiredo, A.R.; Campos, F.; de Freitas, V.; Hogg, T.; Couto, J.A. Effect of phenolic aldehydes and flavonoids on growth and inactivation of *Oenococcus oeni* and *Lactobacillus hilgardii*. *Food Microbiol.* **2008**, *25*, 105–112. [CrossRef]
50. Chasseriaud, L.; Krieger-Weber, S.; Déléris-Bou, M.; Sieczkowski, N.; Jourdes, M.; Teissedre, P.L.; Claisse, O.; Lonvaud-Funel, A. Hypotheses on the effects of enological tannins and total red wine phenolic compounds on *Oenococcus oeni*. *Food Microbiol.* **2015**, *52*, 131–137. [CrossRef]
51. Stivala, M.G.; Villecco, M.B.; Enriz, D.; Fernandez, P.A. Effect of phenolic compounds on viability of 2 wine spoilage lactic acid bacteria. A structure-activity relationship study. *Am. J. Enol. Vitic.* **2017**, 1–12. [CrossRef]
52. Reguant, C.; Bordons, A.; Arola, L.; Rozes, N. Influence of phenolic compounds on the physiology of *Oenococcus oeni* from wine. *J. Appl. Microbiol.* **2000**, *88*, 1065–1071. [CrossRef] [PubMed]
53. Garcia-Ruiz, A.; Moreno-Arribas, M.V.; Martin-Alvarez, P.J.; Bartolome, B. Comparative study of the inhibitory effects of wine polyphenols on the growth of enological lactic acid bacteria. *Int. J. Food Microbiol.* **2011**, *145*, 426–431. [CrossRef] [PubMed]
54. Jimeñez, N.; Esteban-Torres, M.; Mancheño, J.M.; De Las Rivas, B.; Muñoz, R. Tannin degradation by a novel tannase enzyme present in some *Lactobacillus plantarum* strains. *Appl. Environ. Microbiol.* **2014**, *80*, 2991–2997. [CrossRef] [PubMed]
55. Terrade, N.; Mira de Orduña, R. Determination of the essential nutrient requirements of wine-related bacteria from the genera *Oenococcus* and *Lactobacillus*. *Int. J. Food Microbiol.* **2009**, *133*, 8–13. [CrossRef]
56. Liburdi, K.; Benucci, L.; Esti, M. Lysozyme in wine: An overview of current and future applications. *Compr. Rev. Food Sci. Food Saf.* **2014**, *13*, 1062–1073. [CrossRef]
57. Kong, M.; Chen, X.G.; Xing, K.; Park, H.J. Antimicrobial properties of chitosan and mode of action: A state of the art review. *Int. J. Food Microbiol.* **2010**, *144*, 51–63. [CrossRef]
58. Bağder Elmacı, S.; Gülör, G.; Tokath, M.; Erten, H.; İşci, A.; Özċelik, F. Effectiveness of chitosan against wine-related microorganisms. *Antonie Van Leeuwenhoek* **2014**. [CrossRef] [PubMed]
59. Nardi, T.; Vagnoli, P.; Minacci, A.; Gautier, S.; Sieczkowski, N. Evaluating the impact of fungal-origin chitosan preparation on *Brettanomyces bruxellensis* in context of wine aging. *Wine Stud.* **2014**, *3*, 13–15. [CrossRef]
60. Schildberger, B. Influence of Specific Botryticides on Malolactic Fermentation. In *Mitteilungen Klosterneuburg, Rebe und Wein, Obstbau und Früchteverwertung*; Wissensbericht; Lehr- und Forschungszentrum für Wein und Ostbau: Klosterneuburg, Austria, 2011; pp. 168–172.
61. Beelman, R.B.; Kunkee, R.E. Inducing co-inoculation malolactic-alcoholic fermentation in red table wines. In *Proceedings of the Australian Society of Viticulture and Oenology Seminar on Malolactic Fermentation*; Australian Wine Research Institute: Urrbrae, South Australia, 1985; pp. 97–112.
62. Ribéreau-Gayon, J.; Peynaud, E.; Ribéreau-Gayon, P.; Sudraud, P. *Sciences et Techniques du Vin*; Dunod: Paris, France, 1975; Volume 2.
63. Krieger, S.; Zapparoli, G.; Veneri, G.; Tosi, E.; Vagnoli, P. Simultaneous and sequential alcoholic and malolactic fermentations: A comparison for Amarone-style wines. *Aust. N. Z. Grapegrow. Winemak.* **2007**, *517*, 71–77.
64. Sieczkowski, N. Maîtrise et intérêt de la co-inoculation levures-bactéries. *Revue Française D'Oenologie* **2004**, *207*, 24–28.
65. Gerbaux, V.; Briffox, C.; Dumont, A.; Krieger, S. Research Note: Influence of Inoculation with Malolactic Bacteria on Volatile Phenols in Wines. *Am. J. Enol. Vitic.* **2009**, *60–62*, 233–235.
66. Kunkee, R.E. Malolactic fermentation. *Adv. Appl. Microbiol* **1967**, *9*, 235–279.
67. Silva, A.; Lambri, M.; Fumi, M.D. Ochratoxin: A Decontamination by Lactic Acid Bacteria in Wine: Adsorption or Biodegradation? In Proceedings of the Oeno 2007 VIII Symposium International d'Oenologie—Bordeaux, Paris, France, 25–27 June 2007; pp. 24–27.

68. Iorizzo, M.; Testa, B.; Lombardi, A.J.; García-Ruiz, A.; Muñoz-González, C.; Bartolomé, B.; Moreno-Arribas, M.V. Selection and technological potential of Lactobacillus plantarum bacteria suitable for wine malolactic fermentation and grape aroma release. *LWT Food Sci. Technol.* **2016**, *73*, 557–566. [CrossRef]
69. Capozzi, V.; Russo, P.; Ladero, V.; Fernández, M.; Fiocco, D.; Alvarez, M.A.; Grieco, F.; Spano, G. Biogenic amines degradation by *Lactobacillus plantarum*: Toward a potential application in wine. *Front. Microbiol.* **2012**, *3*, 1–6. [CrossRef]
70. Knoll, C.; Divol, B.; Du Toit, M. Genetic screening of lactic acid bacteria of oenological origin for bacteriocin-encoding genes. *Food Microbiol.* **2008**, *25*, 983–991. [CrossRef]
71. Lerm, E.; Engelbrecht, L.; du Toit, M. Selection and Characterisation of *Oenococcus oeni* and *Lactobacillus plantarum* South African Wine Isolates for Use as Malolactic Fermentation Starter Cultures. *S. Afr. J. Enol. Vitic.* **2011**, *32*, 280–295. [CrossRef]
72. Gerbaux, V.; Briffox, C. Influence de l'ensemencement en bactéries lactiques sur l'évolution de la couleur des vins de Pinot noir pendant l'élevage. *Revue des Œnologues* **2003**, *103*, 19–23.
73. Asenstorfer, R.E.; Markides, A.J.; Iland, P.J.; Jones, G.P. Formation of vitisin A during red wine fermentation and maturation. *Aust. J. Grape Wine Res.* **2003**, *9*, 40–46. [CrossRef]
74. Burns, T.R.; Osborne, J.P. Loss of Pinot noir wine color and polymeric pigment after malolactic fermentation and potential causes. *Am. J. Enol. Vitic.* **2015**, *66*, 130–137. [CrossRef]
75. Bartowsky, E.; Krieger-Weber, S. Management of malolactic fermentation to enhance red wine colour. *Grapegrow. Winemak.* **2020**, *673*, 66–70.
76. Mtshali, P.S.; Divol, B.T.; Van Rensburg, P.; Du Toit, M. Genetic screening of wine-related enzymes in Lactobacillus species isolated from South African wines. *J. Appl. Microbiol.* **2010**, *108*, 1389–1397. [CrossRef]
77. Spano, G.; Rinaldi, A.; Ugliano, M.; Moio, L.; Beneduce, L.; Massa, S. A β-glucosidase gene isolated from wine *Lactobacillus plantarum* is regulated by abiotic stresses. *J. Appl. Microbiol.* **2005**, *98*, 855–861. [CrossRef]
78. Lucio Costa, O. Acidificón Biológica de Vinos de pH Elevado Mediante la Utilización de Bacterias Lácticas. Ph.D. Thesis, University of Valencia, Valencia, Spain, 2014.
79. Lucio, O.; Pardo, I.; Krieger-Weber, S.; Heras, H.M.; Ferrer, S. Selection of *Lactobacillus* strains to induce biological acidification in low acidity wines. *LWT Food Sci. Technol.* **2016**, *73*, 334–341. [CrossRef]
80. Saerens, S.; Edwards, N.; Soerensen, K.; Badaki, M.; Swieger, J.H. Production of Low Alcohol Fruit Beverage. International Patent Application Number PCT/EP2015/051163, 21 January 2015.
81. Edwards, N.; Saerens, S.; Swieger, J.H. The Use of Lactobacillus plantarum As an Anti-Microbial Agent in the Process of Winemaking. International Patent Application Number PCT/EP2015/051161, 21 January 2015.

Article

Statistical Modelization of the Descriptor "Minerality" Based on the Sensory Properties and Chemical Composition of Wine

Elvira Zaldívar Santamaría [1,*], David Molina Dagá [2] and Antonio T. Palacios García [1,3]

[1] Laboratorios Excell Ibérica S.L., Planillo 12, pabellón B, 26006 Logroño, La Rioja, Spain; atpalacios@telefonica.net

[2] Outlook Wine, Cinema Bel, 28, Esc C, 2°, 1ª. 08940 Cornella de Llobregat, Spain; dmolina@outlookwine.com

[3] Food Technology Department, University of La Rioja, Avda. de la Paz, 93, 26006 Logroño, La Rioja, Spain

* Correspondence: controlcalidad@labexcell.es; Tel.: +34-675-046290

Received: 31 July 2019; Accepted: 14 November 2019; Published: 22 November 2019

Abstract: When speaking of "minerality" in wines, it is common to find descriptive terms in the vocabulary of wine tasters such as flint, match smoke, kerosene, rubber eraser, slate, granite, limestone, earthy, tar, charcoal, graphite, rock dust, wet stones, salty, metallic, steel, ferrous, etc. These are just a few of the descriptors that are commonly found in the tasting notes of wines that show this sensory profile. However, not all wines show this mineral trace at the aromatic and gustatory level. This study has used the statistical tool partial least squares regression (PLS) to mathematically model the attribute of "minerality" of wine, thereby obtaining formulas where the chemical composition and sensory attributes act jointly as the predictor variables, both for white wines and red wines, so as to help understand the term and to devise a winemaking approach able to endow wines with this attribute if desired.

Keywords: minerality; partial least squares regression; predictive model; white wine; red wine

---

1. Introduction

Certain varieties of grape are more likely than others to generate wine "minerality" imprint, such as the internationally known whites Riesling, Chardonnay, Chenin Blanc, Sauvignon Blanc, Grüner Veltliner and Albariño, among others, and Syrah and Carignan red wines and, to a lesser extent, Cabernet Franc, Merlot, Cabernet Sauvignon and others such as Nebbiolo and Barbera. Among all these wines, some common aspects can be found when they express "minerality": when they are grown in a cold environment and/or marginal climates, harvested in early vintages avoiding over-ripeness, have high acidity or are made through reductive winemaking with a generous dose of sulfur dioxide. They generally tend to be wines with a "single vineyard" profile, potentially seeking to reflect the expression of a terroir. However, it is not exclusive to these "cuvées", since nowadays there are mass production wines in the market at popular prices, produced in different countries from around the world that also reveal a sensory profile with a mineral character. In many cases, they tend to be dry white wines with high acidity and a relevant low fruity aromatic profile. In most cases, the leading market influencers and consumers interpret this perception as a value of intangible quality that praises the hedonic and economic value of the wine [1,2].

There is no doubt that the quality that the term "minerality" in wines transmits is certainly one of the most mysterious attributes from the chemical and sensory point of view. Little was known until now, since no thorough studies had been made before on how certain chemical compounds can affect the description of the term "minerality" by the taster and the consumer. As previously mentioned, "minerality" in wines is often associated with the "terroir" concept, often with clear commercial purposes where the expression linked to the soil allows you to justify or argue the authenticity of the wine's origin, with examples of labels in the market that clearly convey this message with associative images and names. It would therefore be easy to link the term "minerality" to the composition and content of minerals that are present in a wine, even though there are no scientific studies to support this direct association [3].

This study is the corollary of a prior, already published research paper, "Chemical basis of mineral character at olfactory and gustatory level in white and red wines", and it aims to verify the hypothesis that certain chemicals and not essentially the metal content are responsible for the use of the attribute "minerality" in wine. This paper concludes by mentioning the chemicals associated with the term "minerality" and proposes predictive mathematical algorithms against the renowned term.

It is widely known that there is a huge list of descriptors in the wine world to articulate the qualities, types, and styles of wines at the sensory level. Undoubtedly, the use of the term "minerality" has become very popular in the 21st century and is much used by producers, distributors, and particularly by tasters and famous gurus as a relevant indicator of difference and distinction between wines, especially among high-end and high price labels. Referring to "minerality" in the description of a wine entails endowing it with greater potential sensory and commercial value.

Over the last decade, the impact of the interpretation of this term has become internationally important. There is a strong need to find the possible causes and the origin of the association of the term "minerality" with the presence of odoriferous volatile compounds, certain minerals or other aromatic or sapid substances that may come from the soil, the metabolism of the plant itself or as a result of enological treatments applied in the winery.

The lack of a clear, well-argued definition of the term "minerality" has itself become a drawback of this powerful term. This has given rise to the enigmatic division between those who define themselves as "mineralist" who often match the "pro-terroir" profile, and those who are defined as "anti-mineralist", who, in turn, also tend to be skeptical about the very concept of "terroir", perhaps the most powerful marketing term for the wine industry.

The final aim of this study is to examine the possible association of the chemical composition of wine and its sensory attributes with the "minerality" of wine, and to use these elements as variables that are part of predictive mathematical formulas for evaluating the potential mineral character of the wine.

2. Materials and Methods

2.1. Sample Characterization

Seventeen commercial wines were used, including white and red wines from different vintages and worldwide winemaking regions. For the selection of the wines, a specialized press investigation was carried out in order to choose wines that would have been described as "mineral". For this, reviews sourced from specialized magazines (The Wine Advocate, Wines and Spirits, Wines & Vines, etc.) were used. Similarly, preference was given to those wine regions identified as "mineral" regions by specialized magazines from the sector (http://www.thewinesociety.com/society-news-and-views-regular-features-opinion-mineral-wines).

Eleven white wines and six red wines were chosen. The description of the wines used in the study is detailed in Table 1.

Table 1. Identification of the samples of white and red wine used in the study.

N°	Wine Type	Variety	Enological Practices	Year of Harvesting	Region
1	White wine	Godello	No MLF */OBA **	2011	Valdeorras (Spain)
2	White wine	Sauvignon blanc	No MLF */OBA **	2008	Loire Valley (France)
3	White wine	Treixadura	No MLF */OBA **	2011	Ribeiro (Spain)
4	White wine	Godello	No MLF */OBA **	2011	Ribera Sacra (Spain)
5	White wine	Riesling	No MLF */OBA **	2008	Niederosterreich (Austria)
6	White wine	Garnacha Gris	No MLF */OBA **	2011	Empordá (Spain)
7	White wine	Ribolla	No MLF */OBA **	2010	Primoska (Slovenia)
8	White wine	Xarel.lo	No MLF */OBA **	2011	Penedés (Spain)
9	White wine	Riesling	No MLF */OBA **	2010	Mosel (Germany)
10	White wine	Riesling	No MLF */OBA **	2009	Mosel Trocken (Germany)
11	White wine	Riesling	No MLF */OBA **	2009	Mosel Kabinett (Germany)
12	Red wine	Tinta del País	MLF */OBA **	2007	Underwater wine (Spain)
13	Red wine	Blaufrankisch	MLF */OBA **	2008	Burgenland (Austria)
14	Red wine	Syrah	MLF */OBA **	2008	North Rhone Valley (France)
15	Red wine	Poulsard	MLF */OBA **	2010	Jura (France)
16	Red wine	Garnacha, Syrah	MLF */OBA **	2011	Montsant (Spain)
17	Red wine	Syrah	MLF */OBA **	2007	Aragón (Spain)

* MLF: malolactic fermentation; ** OBA: Oak Barrel Aging.

2.2. Sensory Analysis

In order to carry out the sensory part of the study, two tasting panels were trained in accordance with the Asociación Española de Estandarización (UNE) 87024-2. One of them, located in Barcelona, formed by 10 professional judges from the wine export sector not specialized in winemaking, and a second panel located in La Rioja formed by 12 students of oenology. The panels were recruited in two ways: the panel from the University of La Rioja was trained for one year by following the program of the sensory analysis course that is part of the University's degree in Enology. The panel of winemakers from La Rioja was made up of sensory judges aged between 22 and 36 (61% women and 39% men). In the case of the panel of exporters in Barcelona, this was composed of sensory judges aged between 35 and 48 (72% men and 28% women).

The judges from the University of La Rioja were trained with chemical standards and real samples for one year. The second panel, from Barcelona, was made up of students who passed Level 2 of the Wine and Spirits Trust beverages program that the company Outlook Wine S.L. teaches in Barcelona.

Sensory descriptive analysis was performed according to the International Organization for Standardization (ISO) 11035. The wines were served in certified glasses according to ISO 3591-1977, with 50 mL in each glass, covered with a Petri dish so that the aromas reached equilibrium in the head space. The judges were first asked to evaluate the aromas orthonasally and record their intensity for each of the descriptors set out in the tasting sheet, on a scale of 0 to 5 for positive attributes and 0 to −5 for negative attributes or defects The judges were then asked to conduct the evaluation at the gustatory level with the same scale.

2.3. Statistical Analysis

A set of 17 wines were tasted and, subsequently, a partial least squares regression analysis was conducted with data obtained by using XLSTAT 2017 Addinsoft statistical software (40, rue Damrémont 75018 Paris, France).

Partial least squares regression statistical methodology is an analysis that is related to the regression of major components: Instead of finding hyperplanes of maximum variance between the response variable and independent variables, there is a linear regression through the projection of forecast variables and the observable variables to a new space.

In a first step, and in order to reduce the variables that would be part of the predictive mathematical algorithms, various selection criteria were taken into account [4,5].

High level of positive or negative correlation between the chemical compound and the "minerality" attribute and likewise for organoleptic attributes [6].

Percentage of explanatory variance by the model and Q^2 value accumulated as high and as close to 1 as possible.

Subsequently, a dynamic search process for the best PLS model was performed. For this, those compounds and chemical families with a relationship greater than 2 in their concentration between the maximum and minimum value and with activity aroma values (OAV) that met the condition OAVmax/OAVmin >1 were taken into consideration, since these are compounds that can contribute with significant differences in sensory perception against the scores of the "minerality" attribute by the panel of tasters. To reduce the number of variables of the models, a partial least squares (PLS) regression analysis was performed, retaining those chemical compounds that showed a significant correlation with a confidence level of 90% with the olfactory and gustatory "minerality" attribute.

2.4. Chemical Characterization

2.4.1. Volatile Composition

The minor compounds were analyzed by solid phase extraction (SPE) followed by analysis via Gas Chromatography and Mass Spectrometry (GC–MS) and major compounds by liquid–liquid extraction (LL) followed by Gas Chromatography (GC) with flame ionization detection (FID) [7]. To perform the analysis, 5 mL of wine, 0.8 mL of acetonitrile, 0.1 mL of chloroform and an internal standard containing the deuterated compounds 3-octanol, 4-methyl-pentanol and 4-hydroxy-4-methyl-2-pentanol were added to a centrifuge tube with a 15 mL thread. The tube was centrifuged at 4800 rpm for 5 min. Once the phases were separated, the organic phase was recovered and injected into an Agilent gas chromatograph model 7890 (G3440A) under the following conditions: an initial temperature of 40 °C, which was maintained for 5 min and then increased by 3 °C/min to 200 °C. The gas used was Helium injected at a flow of 3 mL/min. Three milliliters were injected with a divided mode flow of 30 mL/min. The column (50 m × 0.32 mm) and film thickness of 0.5 μm used was DB-FFAP (nitroterephthalic-acid-modified polyethylene glycol (PEG) column of high polarity) from J & W Scientific (Folsom, CA, USA) and the identification and quantification was carried out by means of a mass detector. To perform the quantification of the majority and minority compounds, calibration graphs for 70 compounds were previously prepared by analyzing synthetic samples containing known quantities of the odorants analyzed.

2.4.2. Enological Parameters

The quantitative determination of these parameters was carried out following the guidelines of the official methods published in Boletín Oficial del Estado (BOE) number 1988-11256 and the International Organization of Vine and Wine (OIV).

The ethanol content of the samples was determined by near infrared (NIR, Anton Paar; Camino de la Fuente de la Mora, 9, 28050 Madrid, Spain) equipment. The values of pH and total acidity were determined by pH apparatus coupled to an automatic titrator (Mettler-Toledo S.A.E. Miguel Hernández 69-71, 08908 Hospitalet de Llobregat, Barcelona, Spain; 2000 series) calibrated according to certified standards. The parameters referring to the colorant composition of the samples were measured by spectrophotometric techniques using Lan Optics equipment (series 2000).

The amounts of free and total sulfur dioxide were analyzed by steam trawling following the Franz Paul approach and distilling equipment supplied by GAB Technologies. The determination of organic acids and sugar content (glucose + fructose) was based on automated enzymatic methods with measurement by spectrophotometry using enzymatic equipment (Analyzer Y15, Byosistem, Barcelona, Spain). In the case of succinic acid, a manual enzymatic method was followed using the kit supplied by Megazyme (https://www.megazyme.com/) through the generation of calibration and measurement lines by spectrophotometry with Lan Optics equipment (2000 series).

2.4.3. Quantification of Metals

The quantitative determination of metals was made by means of the following methods. A prior digestion was conducted in sealed containers to mineralize the wine samples. The samples were subsequently analyzed by an optical emission spectrometer with excitation by induced argon plasma and a dispersive system as previously described [8]. For the realization of the calibration lines, solutions of each of the reference standard mediating elements supplied by Sigma Aldrich in 5% NO_3H were prepared. The metal quantification analyses were carried out by the regional agricultural laboratory of La Grajera of the Government of La Rioja.

3. Results

3.1. Modelization of Aromatic "Minerality" Based on the Chemical Composition of Wines

For the study of the data obtained by the panel, a principal component analysis (PCA) of each attribute was performed, which was calculated from the average scores of the panel members. A correlation matrix was obtained for each attribute (wines in rows and tasters in columns). Thus, each of the attributes was represented in a factorial plane where the projections of each taster were placed around 360° of the correlation circle. With the average data of the scores established by each panel, a principal component analysis (PCA) was performed to evaluate the results obtained by the panel of non-producing tasting experts.

The PCA of each attribute was calculated with the average scores of the sensory judge members of the panel. For those descriptors in which the judges were grouped in the same part of the plane, it was interpreted as a sign that the panel had the same criteria in the interpretation of the attribute. On the contrary, those attributes in which the PCA showed the projection of the judges distributed throughout the plane could be explained as:

(1) the panelists did not interpret the attribute in the same way.
(2) the sensory differences for that attribute were too small to be perceived by most panelists.

For this reason, only those attributes that showed a PCA with a projection of at least 60% of the judges in the same plane were taken into account for the correlation analysis shown in Table 2.

Table 2. Chemical compounds with correlation at a confidence level of 90% with the olfactory phase (in bold letters with significant differences; * $p < 0.1$; ** $p < 0.05$).

Compound	Correlation Coefficient	Compound	Correlation Coefficient
β-Ionone	−0.147	Ethyl hexanoate	0.177
β-Damascenone	0.121	3-Methylbutanoic acid	−0.364
Butyric acid	−0.407	Ethyl isobutyrate	−0.216
** **Isobutyric acid**	**−0.564**	Ethyl 2-methyl butyrate,	0.042
Hexanoic acid	0.010	Guaiacol	−0.372
* **Octanoic acid**	**0.452**	Eugenol	−0.380
Ethyl isovalerate	0.047	4-ethylphenol	−0.360
** **Phenylethyl alcohol**	**0.558**	γ-decalactone	−0.371
* **Isoamyl acetate**	**−0.464**	2-Methoxy-4-vinylphenol	0.343
Ethyl butyrate	−0.030	2-methyl-3-furanthiol	0.130
** **Ethyl decanoate**	**0.550**	2-furfurylthiol	−0.264
Acetaldehyde	−0.127	** **4-Mercapto-4-methyl-2-pentanone pentanonepentanona**	**0.529**
Ethyl acetate	**−0.543**	3-Mercaptohexyl acetate	−0.385
Diacetyl	−0.169	3-mercaptohexanol	0.364
** **Isoamyl alcohol**	**−0.526**	** **Benzyl mercaptan**	**0.600**

Table 2 describes the results obtained in the correlation study ($p < 0.1$; 90% significance level) among the scores obtained by the panel of tasters, consisting of processors for the olfactory "minerality" descriptor, and volatile and non-volatile chemical compounds from the study samples. In bold are the variables with their correlation coefficients. Only the compounds isobutyric acid, octanoic acid, β-phenylethanol, isoamyl acetate, ethyl acetate, ethyl decanoate, isoamyl alcohol and 4-mercapto-4-4-methyl-2-2-pentanone were shown to be significant for the olfactory "minerality" descriptor, so these are included in the final PLS model. The crucial fact that isobutyric acid, isoamyl acetate and isoamyl alcohol correlate negatively must also be considered.

Next, a correlation study based on the proposed model was conducted for the "minerality" attribute and the variables that show a more positive or negative correlation among the quantified analytes.

Shown below is the proposed mathematical model once the statistical program was applied:

$$\text{Aromatic minerality} = 2.33 - 0.009 \times \text{isobutyric acid} + 0.023 \times \text{octanoic acid} - 0.026 \times \text{isoamyl acetate} - 0.074 \times \text{ethyl acetate} + 0.328 \times \text{ethyl decanoate} - 0.074 \times \text{isoamyl alcohol} + 0.016 \times \text{benzyl mercaptan}.$$

In the graph of quality of fit (Figure 1B), we observe how values of 0.58 of cumulative Q^2 are reached, close therefore to the value of 0.6 required to ensure that the results are interesting. The Q^2 cumulated index measures the overall goodness of fit and the predictive quality of the models proposed.

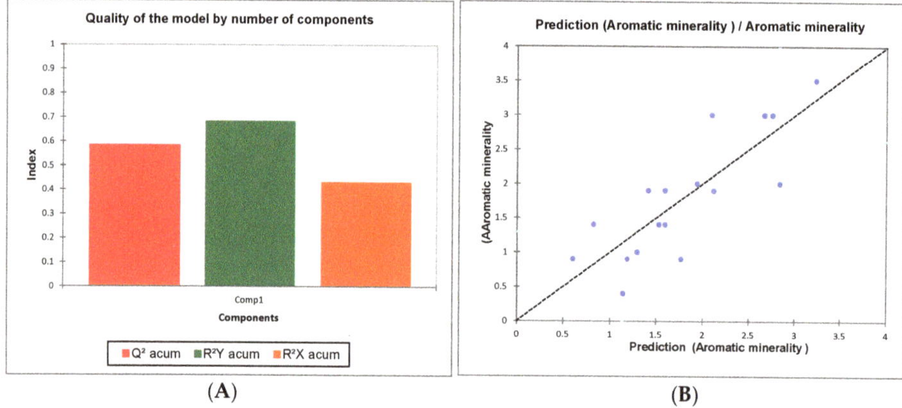

Figure 1. (**A**) Quality of fit of partial least squares (PLS) model; (**B**) Distribution of the seventeen wine samples of the study based on the volatile profile of wines for the olfactory "minerality" descriptor.

The corresponding bar chart shown in Figure 1A enables us to visualize the quality of the partial least squares regression as a function of the number of components.

The closer the cumulated R^2Y and R^2X that correspond to the correlations between the explanatory (X) and dependent (Y) variables and the components are to 1, the better the partial least squares model.

Likewise, the graph of distribution of samples shows how they are projected in a linear way, and how in none of them do the residual values for the model exceed the maximum value of 2 (Figure 1A). These parameters indicate a good fit for the proposed model.

3.2. Modelization of Aromatic "Minerality" Based on the Sensory Properties of Wines

In a similar manner to that followed in the study, a partial least squares (PLS) regression analysis was carried out with data regarding the chemical composition and the volatile fraction of the wines in relation to the aromatic "minerality" attribute, taking into account the sensory attribute scores provided by the panel of expert judges. To choose the descriptors with significant differentiating capacity, those that showed a correlation of up to 90% level of confidence with the olfactory "minerality" descriptor were taken into account. As was stated in the review, it is reported that other descriptors such as granite, limestone, rock, etc., were associated with minerality [9]. However, only "minerality" was included in the sensory analysis so as not to distract the judges from the aim of this study.

The results are described in Table 3. In bold are the descriptors with significant values ($p < 0,1$; 90% significance level) for the "minerality" olfactory descriptor oak, empyreumatic, animal, plant chlorophyll and oxidation, so that these are included in the mathematical predictive model.

Table 3. Sensorial attributes from the sensory analysis performed on seventeen wine samples with correlation at a significant level of 90% with the olfactory "minerality" attribute (in bold letters with significant differences; * $p < 0.1$; ** $p < 0.05$).

Compound	Correlation Coefficient	Compound	Correlation Coefficient
Purity clearness	0.047	* **Empyreumatic**	**0.442**
Floral	0.412	Animal	−0.405
Tropical fruit	0.318	Phenolate	0.401
Ripe fruit	−0.019	Mineral reduction	−0.306
Passion fruit	−0.230	* **Oxidation**	**0.452**
Stone fruit	0.263	Volatile acid	0.301
Patisserie	0.289	* **Plantchlorophyll**	**0.453**
Resin	−0.377	Plant/herbaceous	0.080
** **Oak**	**−0.493**		

Next, on the proposed PLS model, a study of the correlation between the "minerality" olfactory attribute and other sensory descriptors with higher correlation was conducted. The proposed mathematical model is shown here:

$$\text{Aromatic minerality} = 2.32 - 0.39 \times \text{oak} + 0.28 \times \text{empyreumatic} + 0.483 \times \text{plant chlorophyll} + 0.26 \times \text{oxidation}.$$

3.3. Modelization of Aromatic "Minerality" Based on the Chemical Composition and the Sensory Properties of Wines

A third model was elaborated taking into account chemical data with significant correlation ($p < 0.1$, 90%) with the descriptor of "minerality" descriptor, as well as the scores of from descriptors with significant correlation with the aromatic "minerality" descriptor. The proposed model is presented here:

$$\text{Aromatic minerality} = 3.32 + 0.57 \times \text{plant/chlorophyll} + 0.225 \times \text{oxidation} - 0.015 \times \text{isobutyric acid} + 0.003 \times \text{octanoic acid} - 0.025 \times \text{phenylethyl alcohol} - 0.004 \times \text{isoamyl acetate} - 0.11 \times \text{ethyl acetate} + 0.449 \times \text{ethyl decanoate} - 0.081 \times \text{isoamyl alcohol} + 0.008 \times \text{benzyl mercaptan}.$$

The graph of quality of fit (Figure 2B) shows that Q^2 values of 0.69 are reached. Since a 1.0 Q^2 value is considered an optimal fit, we can judge the proposed model as satisfactory.

In the graphic representation of the samples (Figure 2A), we can see how they are distributed in a linear way, and how in none of them do the residual values found for the model exceed the maximum value of 2. It can also be seen that the residual values found are generally very low.

Figure 3 shows the standardized regression coefficients, in which it can be seen that all the variables of the model are within the recommended range of −1.96 to 1.96. Likewise, it can be observed that, according to this model, the "minerality" olfactory descriptor is positively correlated with the presence of organic acids such as octanoic acid as well as benzyl mercaptan. The model also indicates a relationship between herbaceous notes and oxidation with the "minerality" olfactory descriptor.

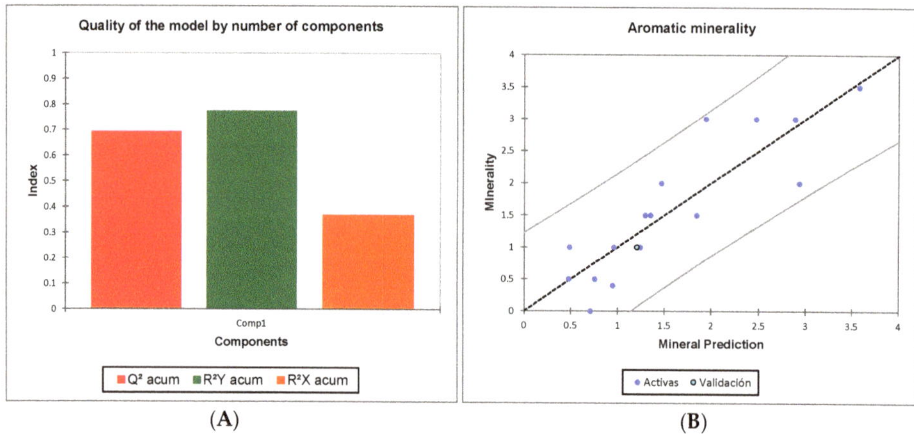

Figure 2. (**A**) Quality of fit of PLS model; (**B**) Distribution of the seventeen wine samples of the study in the predictive model of aromatic "minerality" considering the chemical and sensory profile.

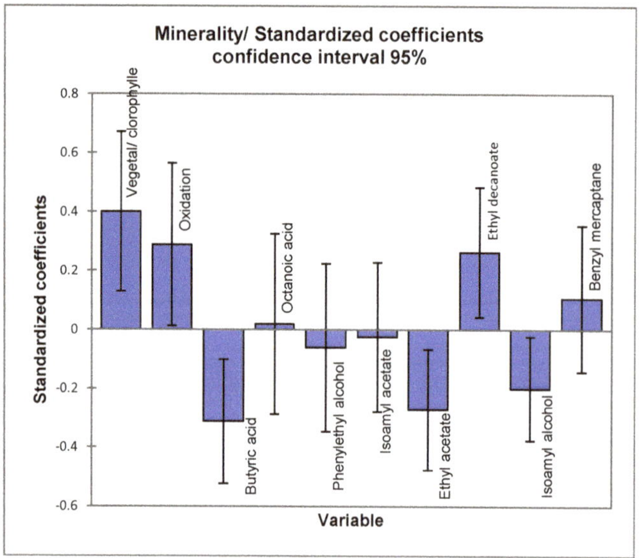

Figure 3. Standardized coefficients of the linear regression in the predictive model of aromatic "minerality" descriptor constructed according to the chemical and sensory profile of the seventeen wine samples of the study.

On the other hand, as noted in the model built based on active olfactory analysis, compounds such as ethyl acetate, nail lacquer aroma and glue, and other compounds such as isoamyl alcohol, fusel or distinctly fruity notes [10,11], as in the case of the banana-scented isoamyl acetate, contribute negatively to the mineral attributes. It is not surprising that the presence of fruity aromas such as those produced by organic esters contribute negatively or contrary to the perception of "minerality", previously described this hypothesis by Par et al. [12–14].

3.4. Modelization of the Gustatory "Minerality"

Similar to what was done with sensory results of the "minerality" attribute at olfactory level, a partial least squares regression (PLS) analysis was conducted with chemical and sensory data regarding gustatory "minerality" obtained by the panel of expert tasters.

Those compounds with a relationship in concentration between the maximum and minimum value greater than 2 were considered, because it is assumed that they are compounds that can make significant differences in gustatory perception. A second criterion was established for this analysis, taking into account only those compounds whose concentration regarding the gustatory sensory threshold was >1. Gustatory thresholds in the case of anions and cations were those considered as known in the water matrix [15], since there are no publications related to wine.

Statistical analyses of the gustatory "minerality" attribute previously analyzed already showed that the results of the ANOVA were significant ($p < 1.35 \times 10^{-2}$) for the set of white and red wines, but less so when the white wines ($p < 8.61 \times 10^{-4}$) and the red wines ($p < 5.61 \times 10^{-2}$) were analyzed separately (data not shown here). Therefore, it was decided to analyze subgroups of white and red wines separately.

3.5. Gustatory "Minerality" Based on the Chemical Composition of White Wines

An initial approach through correlation to a 90% significance level revealed that there was only one compound with a positive correlation between the gustatory "minerality" and the chemical compounds studied. Therefore, it was decided to reduce the level of significance to 60%. A study of correlation was performed for each of the compounds with gustatory capacity, which fulfilled the criteria referred to above, in order to evaluate their discriminatory capacity and include them in the partial least squares regression model. Table 4 describes the results obtained in the correlation study ($p < 0.4$; 60% significance level) of the scores obtained by the tasting panel for the gustatory "minerality" descriptor and the analytical results of compounds related to the gustatory sensations. Descriptors with significant values are highlighted in bold.

Table 4. Chemical compounds from the eleven white wine samples of the study with correlation at a significant confidence level of 60% with the gustatory "minerality" attribute in white wines (in bold letters with significant differences; * $p < 0.4$; ** $p < 0.1$; *** $p < 0.05$).

Compound	Correlation Coefficient	Compound	Correlation Coefficient
Alcoholic strength	−0.254	Glycerol	0.040
Total acidity	−0.236	Aluminum	−0.030
* **Volatile acidity**	**−0.342**	* **Boron**	**−0.330**
** **pH**	**−0.571**	Manganese	0.029
* **L-lactic acid**	**−0.384**	Calcium	0.121
L-malic acid	−0.149	Phosphorus	0.041
Succinic acid	−0.124	* **Magnesium**	**0.461**
** **Tartaric acid**	**0.506**	*** **Potassium**	**−0.595**
* **Glucose & Fructose**	**0.346**		

Variables: volatile acidity, pH, L-lactic acid, tartaric acid, glucose and fructose, boron, magnesium and potassium were shown to be significant, contributing to the definition of the gustatory "minerality" descriptor in white wines, and are therefore included in the model (Figure 4).

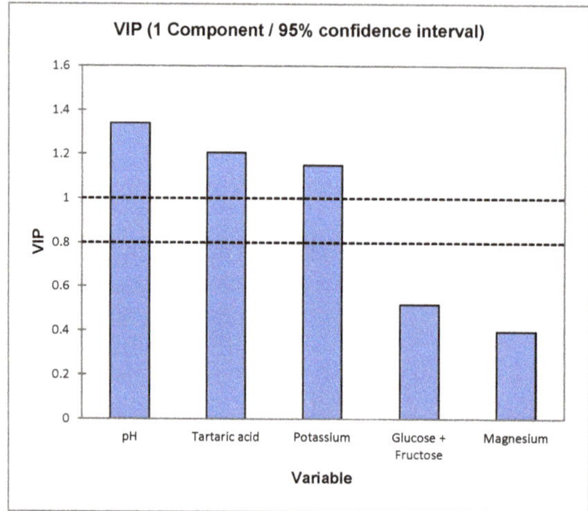

Figure 4. Standardized coefficients of the linear regression model for the gustatory "minerality" descriptor in the eleven white wines of the study constructed on the chemical profile.

Compounds: succinic acid, L-malic acid, alcohol content, total acidity, glycerol, phosphorus, manganese and aluminum were eliminated from the model because its statistical quality improved after removal. Concerning the variable constituted by the sum of glucose and fructose, its coefficient or weight level in formula was very low (0.005), as may be expected, since a priori "minerality" is a difficult character to fit in sweet wines, although in this case we are speaking of dry wines. The proposed model is presented below:

$$\text{Gustatory minerality in white wines} = 5.48 - 1.59 \times \text{pH} + 0.34 \times \text{tartaric acid} + 0.005 \times \text{glucose \& fructose} + 0.011 \times \text{magnesium} - 0.001 \times \text{potassium}.$$

Graph of quality of fit (Figure 5B) shows how the Q^2 value lies in values greater than 0.6, which is the necessary minimum to obtain representative results (0.66). Likewise, none of the wines show residual values greater than 2.0, as shown in the graph of samples dispersion (Figure 5A); therefore, all were retained for the construction of the PLS model.

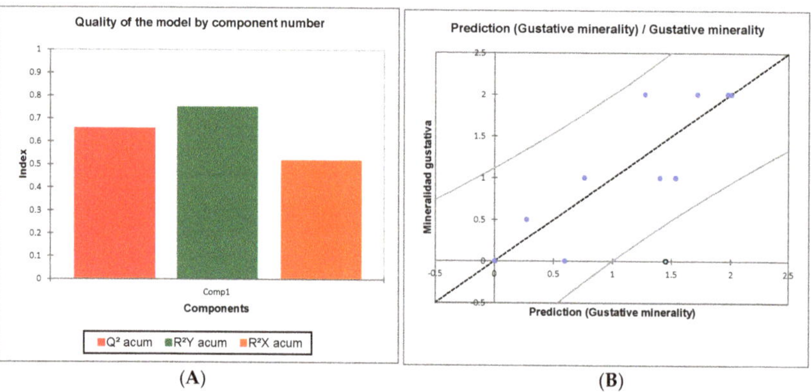

Figure 5. (**A**) Quality of fit of PLS model; (**B**) Distribution of samples of the predictive model for the gustatory "minerality" descriptor based on the chemical profile constructed from the eleven white wines of the study.

It can be noted how the gustatory "minerality" descriptor in white wines was positively related to the increase in acidity. Therefore, the results indicate that increasing levels in tartaric acid and decreasing pH levels favor the emergence of the gustatory "minerality" descriptor in white wines. The model also suggests a positive relationship between the degree of sweetness and "minerality", which is somewhat surprising, but may be due to the fact that one of the wines in the study from the Riesling variety was an "off-dry" style, with 5–7 g/L of residual sugar.

3.6. Gustatory "Minerality" Based on the Sensory Attributes of White Wines

Similarly to what was conducted previously with chemical compounds, a correlation of the gustatory "minerality" attribute with the tasting parameters evaluated in the gustatory phase of the sensory analysis was performed; the results are shown in Table 5 from the scores obtained by the tasting panel for the gustatory "minerality" descriptor and the scores of "minerality". Descriptors with significant values are highlighted in bold.

Table 5. Sensory descriptors with correlation at 60% confidence level with the gustatory "minerality" attribute in the eleven white wines analyzed (in bold letters with significant differences; * $p < 0.4$; ** $p < 0.1$; *** $p < 0.05$).

Compound	Correlation Coefficient	Compound	Correlation Coefficient
* **Sweetness (sugar)**	**0.328**	Tannin (grape origin)	−0.018
* **Level of acidity**	**0.354**	Tannin (oak origin)	−0.026
* **Acidity (freshness)**	**0.343**	Volume (3D sensation)	−0.208
Alcohol (warmth)	−0.145	* **Body (weight)**	**0.392**
*** **Alcohol (sweetness)**	**0.697**	*** **Bitterness**	**0.625**
* **Tannin (concentration)**	**−0.389**	* **Depth**	**−0.354**
Tannin (quality)	−0.132	Gustatory persistence	−0.160
Tannin (astringency)	−0.210	** **Balance**	**−0.428**

Below is the model proposed if taking only into account only the gustatory attributes:

$$\text{Gustatory minerality in white wines} = 0.19 + 0.32 \times \text{sweetness} + 0.234 \times \text{acidity} + 0.25 \times \text{acidity (freshness)} + 0.31 \times \text{alcohol (sweetness)} - 0.30 \times \text{tannin (concentration)} + 0.43 \times \text{body (feeling of weight)} + 0.67 \times \text{bitterness} - 0.31 \times \text{balance}.$$

Again, there is a positive correlation between increasing levels of acidity and the gustatory "minerality" observed by the judges on the panel of tasters. The proposed model reached in this case a value of 0.58 for accumulated index Q^2, which gives the model moderate credibility, since it does not reach the critical value of 0.6.

3.7. Gustatory "Minerality" Based on the Chemical Composition and the Sensory Properties of White Wines

Finally, a third model was developed taking into account both the chemical data with significant correlation ($p < 0.4$, 60%) with the gustatory "minerality" descriptor as well as the descriptors with significant correlation ($p < 0.4$, 60%) with the same descriptor. The proposed model is presented below:

$$\text{Gustatory minerality in white wines} = 2.29 - 0.89 \times \text{pH} + 0.21 \times \text{tartaric acid} + 0.004 \times \text{glucose \& fructose} + 0.014 \times \text{magnesium} - 0.001 \times \text{potassium} + 0.17 \times \text{sweetness} + 0.15 \times \text{acidity level} + 0.20 \times \text{alcohol (sweetness)} + 0.27 \times \text{body} + 0.44 \times \text{bitterness} - 0.19 \times \text{balance}.$$

Fit values found for Q^2 were 0.63 for the first components given that those values greater than 0.6 are acceptable, the proposed model was therefore considered valid.

In addition, and as can be seen in Figure 6, in which the standardized regression coefficients are shown, all the variables of the model are within the recommended range of −1.96 to 1.96. Similarly, it is noted that according to this model, the increasing values of acidity, sweetness and alcohol are positively related to this descriptor. What is more, the absence of balance or the presence of some metals, such as potassium, can contribute negatively to mineral tastes.

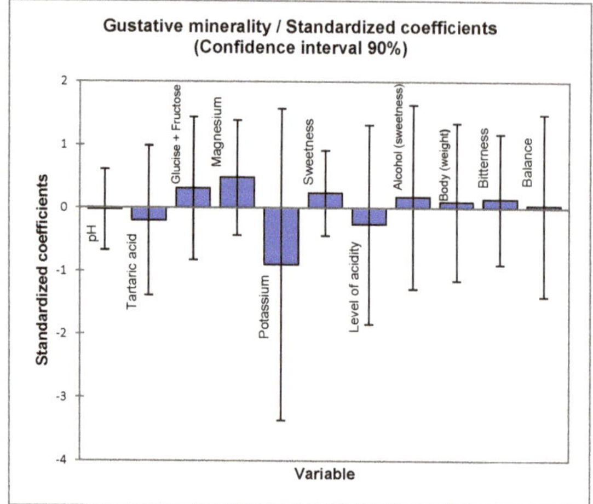

Figure 6. Standardized coefficients of regression for the predictive model of the gustatory "minerality" descriptor based on the chemical and sensory profile from the eleven white wines of the study.

3.8. Modelization of Gustatory "Minerality" Considering the Chemical Composition of Red Wines

As in the white wines, a test was performed on the correlation between the scores of the "minerality" attribute and the concentrations of the different chemical compounds analyzed. The study revealed that there was only one compound with a positive correlation between gustatory "minerality" and the chemical compounds analyzed. Therefore, it was decided to decrease the level of significance to 60%. Only the compounds showing a significant correlation were considered for inclusion in the partial least squares regression model.

Table 6 shows the analytical results of the compounds related to the gustatory sensations. The variables with significant values providing differences between samples are shown in bold. The factors alcoholic strength, L-lactic acid, succinic acid, aluminum, manganese, phosphorus and potassium proved to be significant for the gustatory "minerality" descriptor in red wines and were therefore included in the mathematical model.

Table 6. Chemical compounds with correlation at a significant level of 60% with the gustatory "minerality" attribute in red wines constructed from the data extracted from the six red wines in the study (in bold letters with significant differences; * $p < 0.4$; ** $p < 0.1$).

Compound	Correlation Coefficient	Compound	Correlation Coefficient
* **Alcoholic strength**	**0.505**	* **Aluminum**	**−0.489**
Total acidity	−0.002	Boron	−0.031
pH	0.359	* **Manganese**	**−0.669**
* **L-lactic acid**	**−0.467**	* **Phosphorus**	**0.520**
* **Succinic acid**	**0.637**	** **Potassium**	**0.735**
Glycerol	−0.210		

Below is the proposed PLS model that considers the active chemical composition in the mouth at the sensory level of red wines:

$$\text{Gustatory minerality in red wines} = 0.77 + 0.05 \times \text{alcoholic strength} - 0.15 \times 1 - \text{lactic acid} + 0.99 \times \text{succinic acid} - 0.03 \times \text{aluminum} - 0.17 \times \text{manganese} + 0.001 \times \text{phosphorus} + 0.0003 \times \text{potassium}.$$

3.9. Modelization of Gustatory "Minerality" Considering the Sensory Attributes of Red Wines

Table 7 describes the results obtained in the study of correlation ($p < 0.4$; 60% significance level) of the scores obtained by the panel of expert tasters for the gustatory "minerality" descriptor in red wines

Table 7. Sensory descriptors with correlation at a significant level of 60% with the gustatory "minerality" attribute build from the data extracted from the six red wines of the study (in bold letters with significant differences; * $p < 0.4$; ** $p < 0.1$; *** $p < 0.05$).

Compound	Correlation Coefficient	Compound	Correlation Coefficient
Sweetness (sugar)	0.024	Tannin (grape origin)	−0.202
*** Level of acidity**	**−0.489**	*** Tannin (oak origin)**	**0.489**
*** Acidity (freshness)**	**−0.697**	Volume (3D sensation)	−0.295
**** Alcohol (warmth)**	**0.748**	Body (weight)	−0.259
Alcohol (sweetness)	−0.240	Bitterness	0.352
*** Tannin (concentration)**	**0.697**	*** Depth**	**0.712**
Tannin (quality)	−0.195	**** Taste persistence**	**0.758**
***** Tannin (astringency)**	**−0.847**	Balance	−0.193

The proposed PLS model that considers the attributes of gustative sensory analysis is shown below:

$$\text{Gustatory minerality in red wines} = 1.07 - 0.49 \times \text{acidity (freshness)} + 0.31 \times \text{alcohol (warmth)} + 0.47 \times \text{tannin (concentration)} - 0.34 \times \text{astringency of tannin} + 0.29 \times \text{depth} + 0.24 \times \text{gustatory persistence}.$$

The quality of fit of the two models displayed in Sections 3.1 and 3.2 based on Q^2 parameters were 0.45 for the one which uses chemical parameters and 0.74 for the one which uses sensory attributes; the latter is therefore much more reliable.

In order to improve the fit of the formula that considered the chemical composition, different models were performed eliminating variables whose importance in the projection was less than 0.8 (Figure 7B); however, the model initially proposed for the PLS accumulated the best fit and so it was maintained as valid.

In the sensory model, and as is shown in the Variable Importance in Projection (VIP) graph (Figure 7A), the descriptors tannin from oak and heartburn were eliminated from the model since the statistical quality improved after their removal. Additionally, all the descriptors that are part of the model exceeded the 0.8 cut-off values.

Concerning the chemical composition, it should be highlighted how the gustatory "minerality" descriptor in red wines is positively related to the alcoholic strength and does not seem to give importance to acidity, which was the case in white wines. However, a positive correlation of a compound with a saline character appears in the case of succinic acid. This relationship is already reflected in previous studies about "minerality" in wine.

Furthermore, in relation to the tasting attributes, similarly to what was seen in white wines, there is a positive correlation between the feeling of alcoholic warmth and gustatory "minerality". However, contrary to what happens in white wines, the gustatory acidity in red wines is not decisive in the detection of a mineral character.

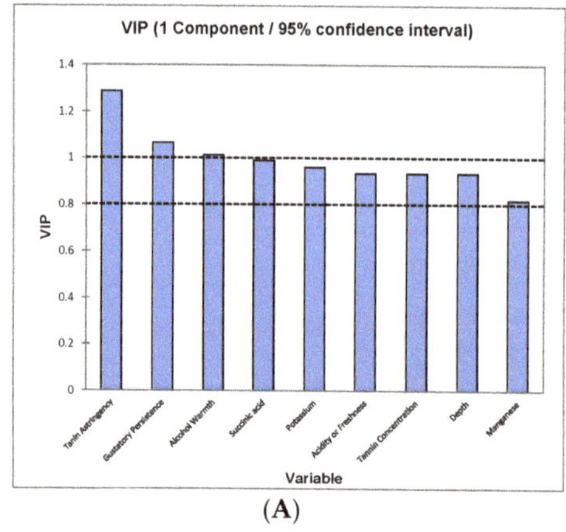

(**A**)

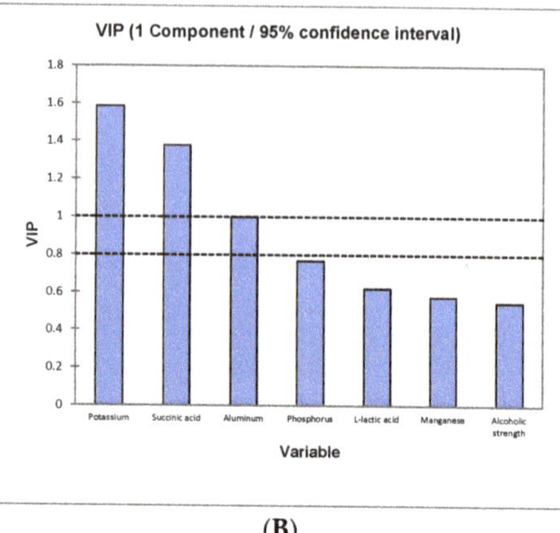

(**B**)

Figure 7. (**A**) Variable Importance in Projection (VIP) graphic of the models proposed for the gustatory "minerality" descriptor in red wines constructed on the chemical profile of active gustatory compounds; (**B**) Idem based on the values of the descriptive sensory analysis from the data extracted from the six red wines in the study.

3.10. Modelization of Gustatory "Minerality" in Red Wines Considering the Chemical Composition and the Sensory Attributes

Finally, a third model was developed in red wines taking into account both chemical data with significant correlation ($p < 0.4$, 60%) with the gustatory "minerality" descriptor and the sensory tasting parameters. The proposed model is presented below:

$$\text{Gustatory minerality in red wines} = 1.84 - 0.29 \times \text{acidity (freshness)} + 0.22 \times \text{alcohol (warmth)} + 0.29 \times \text{tannin (concentration)} - 0.23 \times \text{astringency of tannin} + 0.19 \times \text{depth} + 0.17 \times \text{persistence} + 0.74 \times \text{succinic acid} - 0.25 \times \text{manganese} + 0.0002 \times \text{potassium}.$$

As shown in the graph of variable importance (Figure 8), alcoholic strength, l-lactic acid, aluminum and phosphorus were eliminated from the model since its statistical quality improved after their removal. The model shows a fit of 0.64 based on the accumulated Q^2 parameter, so it can be considered that it exceeds the cutoff threshold established at 0.6.

The model constructed shows how increasing values of alcohol sensations (warmth) and tannins, both at the level of astringency and phenolic concentration, are positively related to this descriptor, and how the model, once again, appears to give no importance to the acidity, as occurred in white wines. However, there is again a positive correlation with a compound with a saline character, succinic acid. The presence of metals is irrelevant, since some of them, such as manganese, contribute positively to the final pls mathematical model and others, like potassium, do so negatively.

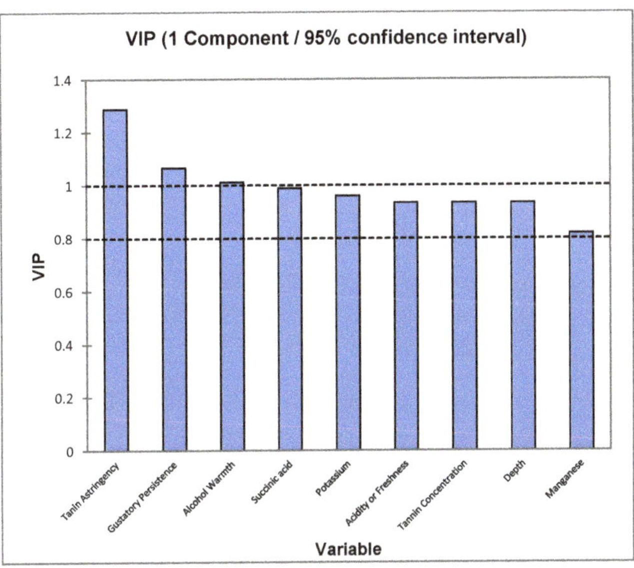

Figure 8. Variable Importance in Projection (VIP) graph of the model proposed for the gustatory "minerality" descriptor in red wines, based on the chemical and sensory profile of the six red wine samples.

4. Discussion

Compared to other studies on "minerality", the present analysis of this descriptor is an innovative approach as it combines sensory and chemical characterization.

Previous studies (followed a purely sensory approach to this aspect of wine. Our proposed models integrate chemical and sensory analysis in order to explain the "minerality" descriptor. Our theory is based on the idea that the role that this descriptor plays in sensory analysis cannot be explained by a unique chemical compound. The close examination of the data collected from the study revealed several models that integrate chemical and sensory descriptors.

Our results obtained after having performed the statistical analysis of the chemical composition with the sensory analysis suggest that some volatile chemical compounds are involved in the application of the term "minerality", the origin of which is defined by the plant's metabolism. However, due to the fermentative activity of yeast and bacteria these are transformed into active compounds from the sensory viewpoint; thus, the "minerality" of wine can also be dependent on winemaking techniques and enological itineraries used in the production of wine in the winery and during the process of wine aging, so it is not solely dependent on agroclimatic factors.

Concerning the olfactory spectrum, and based on the results obtained from the descriptive sensory analysis by the panel of expert tasters, different mathematical algorithms with more than reasonable predictive qualities were developed in the mineral description of wine at the olfactory level, which include the following terms or chemical compounds:

With a positive contribution: plant-chlorophyll, oxidation, octanoic acid, ethyl decanoate, isoamyl alcohol and benzyl mercaptan.

With a negative contribution: isobutyric acid, β-phenylethanol, isoamyl acetate and ethyl acetate.

The analysis of the proposed olfactory model revealed a positive association between oxidation and the term "minerality", suggesting that in those samples with less floral and fresh fruit impact, "minerality" is more likely to appear in a sensory characterization. In addition, there is a negative association with the term "oak descriptors" according to the predictive model developed. In the light of these results, it could be concluded that tertiary flavors, such as those created by the oak barrel aging, are negative, but the oxidation process not related to oak barrel aging is positive.

The mathematical model obtained by partial least squares regression for the gustatory "minerality" in white wines suggested a positive relationship with high total acidity levels. In addition, in terms of elements involved in the mathematical relations, the following compounds should be considered:

With a positive contribution: tartaric acid, glucose, fructose (in the form of residual sugars in dry wines or "off-dry" type), magnesium, sweetness (sugar), level of total acidity, alcohol (sweetness), body and bitterness.

With a negative contribution: pH, potassium and balance in the mouth.

In red wines, the proposed mathematical model achieved good results for accounting for gustatory "minerality". This model is positively related to organic acids, such as succinic acid with a salty taste and tannin concentration, and negatively related to the feeling of freshness and well-integrated or balanced acidity. The factors to be considered in this case are as follows:

With a positive contribution: alcohol (feeling of warmth), tannin concentration, gustatory depth, and persistence in the aftertaste, succinic acid, and potassium.

With a negative contribution: level of acidity in relation to freshness in the mouth, astringency of tannin and manganese.

As a conclusion, the present procedure, based on partial least squares regression, has demonstrated revealing and promising results. Based on these predictive models, it is possible to focus the technology used in the vineyard and the winery to imprint or to increase the mineral character in wine if required for reasons of marketing, product range, corporate communication, competitiveness or market strategy. The same approach is presumably equally applicable to other descriptors in the food sector that are nowadays not well characterized.

Author Contributions: Conceptualization D.M.D. and A.T.P.G.; methodology E.Z.S. and A.T.P.G.; validation E.Z.S. and A.T.P.G.; formal analysis E.Z.S.; investigation E.Z.S., A.T.P.G. and D.M.D.; resources E.Z.S., A.T.P.G. and D.M.D.; data curation A.T.P.G.; writing—original draft preparation E.Z.S.; writing—review and editing A.T.P.G. and D.M.D.; supervision A.T.P.G.

Funding: Laboratorios Excell Iberica S.L. and Outlook Wine S.L.

Conflicts of Interest: The authors declare no conflicts of interest.

References

1. Cattin, P.; Wittink, D.R. Commercial use of conjoint analysis: A survey. *J. Mark.* **1982**, *46*, 44–53. [CrossRef]
2. Moskowitz, H.R.; Silcher, M. The applications of conjoint analysis and their possible uses in sensometrics. *Food Qual. Prefer.* **2006**, *17*, 145–165. [CrossRef]
3. Maltman, A. Minerality in wine: A geological perspective. *J. Wine Res.* **2013**, *24*, 169–181. [CrossRef]
4. Aznar, M.; López, R.; Cacho, J.F.; Ferreira, V. Identification and quantification of impact odorants of aged red wines from Rioja. GC-Olfactometry, quantitative GC-MS, and odor evaluation of HPLC fractions. *J. Agric. Food Chem.* **2001**, *49*, 2924–2929. [CrossRef]
5. Bécue-Bertaut, M.; Pagès, J. Multiple factor analysis and clustering of a mixture of quantitative, categorical and frequency data. *Comput. Stat. Data Anal.* **2008**, *52*, 3255–3268. [CrossRef]
6. Boutou, S.; Chatonnet, P. Rapid headspace solid-phase microextraction/gas chromatographic/mass spectrometric assay for the quantitative determination of some of the main odorants causing off-flavours in wine. *J. Chromatogr. A* **2007**, *1141*, 1–9. [CrossRef]
7. Lopez, R.; Aznar, M.; Cacho, J.; Ferreira, V. Determination of minor and trace volatile compounds in wine by solid-phase extraction and gas chromatography with mass spectrometric detection. *J. Chromatogr. A* **2002**, *966*, 167–177. [CrossRef]
8. Moreno, I.M.; González-Weller, D.; Gutierrez, V.; Marino, M.; Cameán, A.M.; González, A.G.; Hardisson, A. Determination of Al, Ba, Ca, Cu, Fe, K, Mg, Mn, Na, Sr and Zn in red wine samples by inductively coupled plasma optical emission spectroscopy: Evaluation of preliminary sample treatments. *Microchem. J.* **2008**, *88*, 56–61. [CrossRef]

9. Deneulin, P.; Bavaud, F. Analyses of open-ended questions by renormalized associativities and textual networks: A study of perception of minerality in wine. *Food Qual. Prefer.* **1960**, *47*, 34–44. [CrossRef]
10. Escudero, A.; Campo, E.; Fariña, L.; Cacho, J.; Ferreira, V. Analytical characterization of the aroma of five premium red wines. Insights into the role of odor families and the concept of fruitiness of wines. *J. Agric. Food Chem.* **2007**, *55*, 4501–4510. [CrossRef] [PubMed]
11. Escudero, A.; Asensio, E.; Cacho, J.; Ferreira, V. Sensory and chemical changes of young white wines stored under oxygen. An assessment of the role played by aldehydes and some other important odorants. *Food Chem.* **2002**, *77*, 325–331.
12. Catania, C.; Avagnina, S. *Los Aromas Responsables de la Tipicidad y Vinosidad*; Curso Superior de Degustación de Vinos; EEA-INTA: Mendoza, Argentina, 2007.
13. Parr, W.; Maltman, A.; Easton, S.; Ballester, J. Minerality in wine: Towards the reality behind the myths. *Beverages* **2018**, *4*, 77. [CrossRef]
14. Parr, W.V.; Valentin, D.; Breitmeyer, J.; Peyron, D.; Darriet, P.; Sherlock, R.; Ballester, J. Perceived minerality in Sauvignon blanc wine: Chemical reality or cultural construct. *Food Res. Int.* **2016**, *87*, 168–179. [CrossRef] [PubMed]
15. Cohen, J.M.; Kamphake, L.J.; Harris, E.K.; Woodward, R.L. Taste threshold concentrations of metals in drinking water. *J. Am. Water Works Assoc.* **1960**, *52*, 660–670. [CrossRef]

MDPI
St. Alban-Anlage 66
4052 Basel
Switzerland
Tel. +41 61 683 77 34
Fax +41 61 302 89 18
www.mdpi.com

Beverages Editorial Office
E-mail: beverages@mdpi.com
www.mdpi.com/journal/beverages

www.ingramcontent.com/pod-product-compliance
Lightning Source LLC
LaVergne TN
LVHW070636170726
843515LV00004B/257
* 9 7 8 3 0 3 6 5 1 0 9 0 3 *